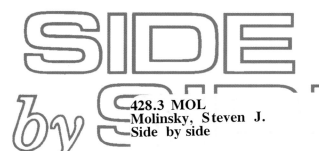

SIDE by SIDE TV

428.3 MOL
Molinsky, Steven J.
Side by side

VIDEO WORKBOOK 1B

Contributing Authors
Elizabeth Handley
Judy Boyle

PRENTICE HALL REGENTS

VIDEO WORKBOOK 1B

Publisher: Tina Carver
Editorial Production/Design Manager: Dominick Mosco
Manufacturing Buyer: Ray Keating
Project Manager: Harriet Dishman
Design and Composition: PC&F, Inc.
Video Stills: Elizabeth Gallagher

Illustrations: Richard E. Hill

Clock times in this workbook enable users to easily find the corresponding video material. Page numbers that appear on the videos refer to pages in the *Side by Side* textbooks, which can also be used as part of this video-based language learning program.

© 1995 by Prentice Hall Regents

In association with

Video Publishing Group, Inc.

All rights reserved. No part of this publication may be reproduced, stored in a retrieval system, or transmitted, in any form or by any means, electronic or mechanical, photocopying, recording, or otherwise, without the prior written permission of the publisher.

Printed in the United States of America

10 9 8 7

ISBN 0-13-158155-4

CONTENTS

PREFACE iv	SEGMENT 21 75
SEGMENT 14 1	SEGMENT 22 87
SEGMENT 15 13	SEGMENT 23 105
SEGMENT 16 29	SEGMENT 24 119
SEGMENT 17 39	SEGMENT 25 131
SEGMENT 18 47	SEGMENT 26 147
SEGMENT 19 57	ANSWER KEY 157
SEGMENT 20 65	

PREFACE

Side by Side TV combines education and entertainment through a variety of comedy sketches, on-location interviews, rap numbers, and music videos—all designed to help learners of English build their language skills. This innovative program may be used in conjunction with the *Side by Side* textbook series, or on its own as an exciting "stand-alone" video-based English course.

Each level of *Side by Side TV* is divided into 13 segments. Within each segment there are short scenes that highlight important structures, functions, and vocabulary items.

The *Side by Side TV Video Workbooks* are designed to be used with the videos as a learning companion. Corresponding to each level of the video, they offer an outstanding variety of exercises and activities to promote language development. The *Video Workbooks* can be used in class or for self-study at home.

FEATURES OF THE VIDEO WORKBOOKS

- *Segment-Opening Pages* indicate the language focus of the segment, the scenes of the segment, and key vocabulary featured in that segment.

- *Exercises and Activities* are intended to help learners interact with the scenes in the video. Certain exercises and activities require use of the video, while others do not. Those that require use of the video are indicated with a small videocassette symbol. For each scene, the clock-time on the video is indicated on the workbook page.

  We encourage viewers to develop their own strategies for working with the exercises. Some viewers may wish to watch a scene one or more times before doing an exercise. Others may wish to complete an exercise while watching a scene. And others may wish to do an exercise before watching, as a way of predicting what might happen in the scene.

- *Scripts* are provided at the end of each workbook segment. Viewers can read along as they watch, read before they watch to preview a scene, or read later for review and practice.

- A *Summary Page* highlights the grammar structures and functions featured in that segment.

The mission of *Side by Side TV* is to provide learners of English with an exciting, motivating, and enjoyable language learning experience through television. We hope that *Side by Side TV* and the accompanying *Video Workbooks* offer viewers a language learning experience that is dynamic, interactive, . . . and fun!

Steven J. Molinsky
Bill Bliss

SEGMENT 14

- Simple Present Tense
- Yes/No Questions
- Negatives
- Short Answers

"For a very special dinner Stanley's Restaurant is a winner . . . eating Side by Side."

PROGRAM LISTINGS

:09 STANLEY'S INTERNATIONAL RESTAURANT
Stanley's Restaurant is a very special place. Every day Stanley cooks a different kind of food.

2:07 WE'RE FILMING A COMMERCIAL
An interviewer talks to passersby about Stanley's International Restaurant.

3:55 A VERY SPECIAL PLACE
An interview with Stanley the chef.

SBS-TV Backstage Bulletin Board

TO: Production Crew

Sets and props for this segment:

Stanley's Kitchen
- pots
- pans
- tablecloth

Reception Desk
- telephone
- flags

Inside Stanley's Restaurant
- tables
- chairs
- dishes
- silverware
- tablecloths

Outside Stanley's Restaurant
- microphone

TO: Cast Members

Key words in this segment:

American
Chinese
Greek
Italian
Japanese
Mexican
Puerto Rican

Monday
Tuesday
Wednesday
Thursday
Friday
Saturday
Sunday

:09 STANLEY'S INTERNATIONAL RESTAURANT

 **SOUND CHECK 1**

① a. (Italian)
　b. Thai

② a. Korean
　b. Greek

③ a. Chinese
　b. Japanese

④ a. Puerto Rican
　b. Portuguese

⑤ a. Japanese
　b. Taiwanese

⑥ a. Moroccan
　b. Mexican

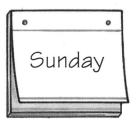

⑦ a. American
　b. Middle Eastern

 SOUND CHECK 2

| cook | cooks | does |

A. Stanley's International Restaurant.

B. What kind of food does Stanley ___cook___¹ on Monday?

A. On Monday he _____² Italian food.

B. Thank you.

SEGMENT 14

A. Stanley's International Restaurant.

B. What kind of food _____³_____ Stanley _____⁴_____ on Wednesday?

A. On Wednesday he _____⁵_____ Chinese food.

B. Chinese food?

A. Yes, that's right.

B. Thank you.

A. Stanley's International Restaurant.

B. _____⁶_____ Stanley _____⁷_____ Greek food on Tuesday?

A. Yes, he _____⁸_____.

B. Thank you.

A. Stanley's Restaurant.

B. Does Stanley _____⁹_____ Puerto Rican food on Thursday?

A. Yes, he _____¹⁰_____.

B. Thanks.

ON CAMERA

You work at Stanley's Restaurant. Complete the following conversations.

A. Good morning. Stanley's Restaurant.

B. Good morning. What kind of food does Stanley cook on _____?

A. On _____ he cooks _____ food.

B. Thank you.

A. Good evening. Stanley's Restaurant.

B. Does Stanley cook _____ food on _____?

A. Yes, he does.

B. Thanks.

SEGMENT 14

SOUND CHECK 3

| doesn't | does | cook | cooks |

A. __Does__¹ Stanley cook Japanese food on Sunday?

B. No, he _____².

A. When _____³ he _____⁴ Japanese food?

B. He _____⁵ Japanese food on Friday.

A. Excuse me?

B. Yes, ma'am.

A. _____⁶ Stanley _____⁷ Chinese food on Monday?

B. No, he _____⁸.

A. When _____⁹ he _____¹⁰ Chinese food?

B. He _____¹¹ Chinese food on Wednesday.

A. I see. Thank you.

ON CAMERA

You're the manager at Stanley's International Restaurant. People are asking you questions about Stanley's menu. Complete these conversations and then practice them with a friend.

A. Excuse me.

B. Yes?

A. Does Stanley cook _____ food on _____?

B. Yes, he does.

A. Oh, great! I like _____ food! See you on _____!

A. Does Stanley cook _____ food on _____?

B. No, he doesn't.

A. I see. Tell me, when does he cook _____ food?

B. He cooks _____ food on _____.

A. Oh. Thanks very much.

2:07 WE'RE FILMING A COMMERCIAL

SOUND CHECK

| cook | go | doesn't | like | do | don't |

A. ___Do___¹ you _____² to Stanley's Restaurant on Wednesday?

B. Yes, I _____³.

A. Why?

B. Because I _____⁴ Chinese food.

A. Thank you very much, sir.

B. My pleasure.

A. _____⁵ you _____⁶ to Stanley's Restaurant on Sunday?

B. No, I _____⁷.

A. Why not?

B. Because I _____ _____⁸ American food.

A. I see. Well, thank you anyway.

A. What kind of food _____⁹ you _____¹⁰?

B. I _____¹¹ Russian food.

A. When _____¹² you _____¹³ to Stanley's Restaurant?

B. I _____¹⁴ go there.

A. Why not?

B. Because Stanley _____ _____¹⁵ Russian food.

SEGMENT 14

ON CAMERA

You're filming a commercial for Stanley's Restaurant. Complete these interviews and then practice them.

Daily Specials

| Monday | Tuesday | Wednesday | Thursday | Friday | Saturday | Sunday |
| Italian | Greek | Chinese | Puerto Rican | Japanese | Mexican | American |

Commercial #1
Location: Outside Stanley's Restaurant

YOU: Excuse me. We're filming a commercial for Stanley's Restaurant. May I ask you a few questions?

PERSON ON STREET: Of course.

YOU: Do you go to Stanley's Restaurant on ....?

PERSON ON STREET: Yes, I do.

YOU: Why?

PERSON ON STREET: Because I like .... food.

YOU: Thank you.

PERSON ON STREET: My pleasure.

Commercial #2
Location: Inside Stanley's Restaurant

YOU: Excuse me. We're filming a commercial. May I ask you a few questions?

PERSON IN RESTAURANT: Certainly.

YOU: Do you go to Stanley's Restaurant on ....?

PERSON IN RESTAURANT: No, I don't.

YOU: You don't? Why not?

PERSON IN RESTAURANT: Because on .... he cooks .... food, and I don't like .... food.

YOU: I see. Thank you anyway.

SEGMENT 14

Commercial #3
Location: Outside Stanley's Restaurant

YOU: Excuse me. We're filming a television commercial. May I ask you a few questions?

PERSON ON STREET: Sure.

YOU: Tell me, what kind of food do you like?

PERSON ON STREET: I like _____ food.

YOU: _____ food?

PERSON ON STREET: Yes, that's right.

YOU: When do you go to Stanley's Restaurant?

PERSON ON STREET: I don't go there.

YOU: Why not?

PERSON ON STREET: Because Stanley doesn't cook _____ food.

YOU: He doesn't?

PERSON ON STREET: No, he doesn't.

SCRAMBLED SOUND TRACK

The sound track is all mixed up. Put the words in the correct order.

1. to | on | Do | Restaurant | Tuesday | you | Stanley's | ? | go
   Do you go to Stanley's Restaurant on Tuesday?

2. you | kind | What | do | of | ? | like | food
   _____

3. French | doesn't | . | Stanley | cook | food
   _____

4. Friday | ? | cook | Does | food | on | Chinese | Stanley
   _____

5. because | Sunday | Restaurant | I | American | to | like | go | food | . | on | don't | Stanley's | I | don't
   _____

SEGMENT 14

3:55 A VERY SPECIAL PLACE

EDITING MIX-UP

The video editor made a mistake! Put the following lines in the correct order.

_____ Thank you.

_____ speak Greek,

_____ speak Italian,

_____ eat Greek food,

_____ drink Italian wine,

_____ and listen to Greek music.

_____ On Monday they

_____ drink Greek wine,

_____ On Tuesday they

_____ and listen to Italian music.

_____ What do people do at Stanley's International Restaurant?

_____ eat Italian food,

__1__ Stanley, your restaurant is a very special place.

STANLEY'S FAVORITE CUSTOMERS

1 Maria likes Chinese food. When does she go to Stanley's Restaurant?

_____

What does she do there?

_____

_____

2 George likes Mexican food. When does he go to Stanley's Restaurant?

_____

What does he do there?

_____

_____

3 Mr. and Mrs. Wong like Greek food. When do they go to Stanley's Restaurant?

_____

What do they do there?

_____

_____

SEGMENT 14

WHAT'S MY LINE?

1. (like) I _____like_____ American food. My wife _____ Greek food.
2. (work) Barbara _____ at the bank. Her brother _____ at the post office.
3. (listen) We _____ to the radio in our car. Do you _____ to the radio in your car?
4. (study) I _____ French, and my sister _____ German.
5. (go) We _____ to Stanley's Restaurant on Tuesday. Our next-door neighbor _____ there on Friday.
6. (speak) The students in this class _____ many different languages. Kenji _____ Japanese. Kim _____ Korean. Manuel and Elena _____ Spanish.
7. (cook) Mr. Lane _____ dinner on Monday. Mrs. Lane _____ dinner on Wednesday. Their son _____ dinner on Friday.

DO THEY OR DON'T THEY?

do	does	don't	doesn't

1. Do you exercise every day? Yes, I _____do_____.
2. Does your daughter study every night? Yes, she _____.
3. Does Mr. Chang work here? No, he _____.
4. Do your children like their new school? Yes, they _____.
5. Does your boss speak Spanish? Yes, he _____.
6. Do you like American food? No, we _____.
7. Do Mr. and Mrs. Ford live nearby? No, they _____.
8. Do I ask a lot of questions? Yes, you _____.
9. Does Betty play the guitar? No, she _____.
10. Do you work on Saturday? No, I _____.

SEGMENT 14 SCRIPT

:09 STANLEY'S INTERNATIONAL RESTAURANT

ANNOUNCER: Stanley's International Restaurant is a very special place. Every day Stanley cooks a different kind of food. On Monday he cooks Italian food. On Tuesday he cooks Greek food. On Wednesday he cooks Chinese food. On Thursday he cooks Puerto Rican food. On Friday he cooks Japanese food. On Saturday he cooks Mexican food. And on Sunday he cooks American food.

(The cashier answers the phone.)

CASHIER: Stanley's International Restaurant.
CALLER 1: What kind of food does Stanley cook on Monday?
CASHIER: On Monday he cooks Italian food.
CALLER 1: Thank you.

CASHIER: Stanley's International Restaurant.
CALLER 2: What kind of food does Stanley cook on Wednesday?
CASHIER: On Wednesday he cooks Chinese food.
CALLER 2: Chinese food?
CASHIER: Yes, that's right.
CALLER 2: Thank you.

CASHIER: Stanley's International Restaurant.
CALLER 3: Does Stanley cook Greek food on Tuesday?
CASHIER: Yes, he does.
CALLER 3: Thank you.

CASHIER: Stanley's Restaurant.
CALLER 4: Does Stanley cook Puerto Rican food on Thursday?
CASHIER: Yes, he does.
CALLER 4: Thanks.

CUSTOMER 1: Does Stanley cook Japanese food on Sunday?
WAITER: No, he doesn't.
CUSTOMER 1: When does he cook Japanese food?
WAITER: He cooks Japanese food on Friday.

CUSTOMER 2: Excuse me?
WAITER: Yes, ma'am.
CUSTOMER 2: I have a question. Does Stanley cook Chinese food on Monday?
WAITER: No, he doesn't.
CUSTOMER 2: When does he cook Chinese food?
WAITER: He cooks Chinese food on Wednesday.
CUSTOMER 2: I see. Thank you.

2:07 WE'RE FILMING A COMMERCIAL

INTERVIEWER: Excuse me, sir. We're filming a commercial for Stanley's Restaurant. May I ask you one

	or two questions?
MAN 1:	Certainly. Go right ahead.
CAMERAMAN:	Rolling!
INTERVIEWER:	Do you go to Stanley's Restaurant on Wednesday?
MAN 1:	Yes, I do.
INTERVIEWER:	Why?
MAN 1:	Because I like Chinese food.
INTERVIEWER:	Thank you very much, sir.
MAN 1:	My pleasure.
INTERVIEWER:	Excuse me. We're filming a commercial.
WOMAN:	A commercial? You mean . . . a TV commercial?!
INTERVIEWER:	Yes. May I ask you one or two questions?
WOMAN:	Sure.
CAMERAMAN:	Rolling!
INTERVIEWER:	Do you go to Stanley's Restaurant on Sunday?
WOMAN:	No, I don't.
INTERVIEWER:	Why not?
WOMAN:	Because I don't like American food.
INTERVIEWER:	I see. Well, thank you anyway.
INTERVIEWER:	Excuse me, sir. We're doing a commercial. May I ask you a few questions?
MAN 2:	Sure.
INTERVIEWER:	What kind of food do you like?
MAN 2:	I like Russian food.
INTERVIEWER:	When do you go to Stanley's Restaurant?
MAN 2:	I don't go there.
INTERVIEWER:	Why not?
MAN 2:	Because Stanley doesn't cook Russian food.
INTERVIEWER:	Cut!

(To the cameraman.)

　　　　　Stanley doesn't cook Russian food?
CAMERAMAN:	No, he doesn't.
MAN 2:	No, he doesn't.
INTERVIEWER:	Thanks anyway.
MAN 2:	No problem.

(To the cameraman.)

| INTERVIEWER: | Why didn't you tell me he doesn't cook Russian food? |

3:55　A VERY SPECIAL PLACE

INTERVIEWER:	Stanley, your restaurant is a very special place.
STANLEY:	Thank you.
INTERVIEWER:	What do people do at Stanley's International Restaurant?
STANLEY:	On Monday, they speak Italian, eat Italian food, drink Italian wine, and listen to Italian music. On Tuesday, they speak Greek, eat Greek food, drink Greek wine, and listen to Greek music. On Wednesday, they speak Chinese, eat Chinese food, drink Chinese…
INTERVIEWER:	Thank you, Stanley. Yes, Stanley's International Restaurant is a very special place. So come on over to Stanley's Restaurant, the International Restaurant, where Stanley cooks a different kind of food every day. Okay?
CAMERAMAN:	It's a wrap.
STANLEY:	It's a wrap?!
INTERVIEWER:	Yes. We're finished.
STANLEY:	It's a wrap!

SEGMENT 14 SUMMARY

GRAMMAR

Simple Present Tense: Yes/No Questions

Do	I / we / you / they	go to Stanley's Restaurant?
Does	he / she / it	

Negatives

I / We / You / They	don't	like American food.
He / She / It	doesn't	

Short Answers

Yes,	I / we / you / they	do.
	he / she / it	does.

No,	I / we / you / they	don't.
	he / she / it	doesn't.

FUNCTIONS

Asking for and Reporting Information

What *do you do there?*
What kind of *food does Stanley cook on Monday?*
When *does he cook Japanese food?*

Do you *go to Stanley's Restaurant?*
    Yes, I do.
    No, I don't.

Does *Stanley cook Greek food on Tuesday?*
    Yes, he does.
    No, he doesn't.

May I ask you *a few questions?*

Inquiring about Likes/Dislikes

What kind of *food* do you like?

Expressing Likes

I like *Chinese food.*

Expressing Dislikes

I don't like *American food.*

Checking Understanding

Chinese food?

Attracting Attention

Excuse me.

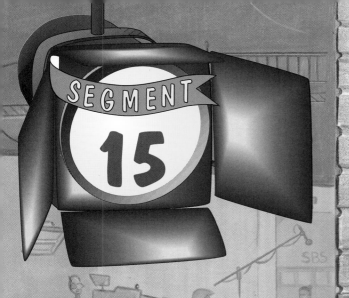

SEGMENT 15

- **Favorite Types of Entertainment**
- **Simple Present Tense**

"Our favorite sport, our favorite book, our favorite food we like to cook . . . together Side by Side."

PROGRAM LISTINGS

5:04 SBS-TV ON LOCATION
People tell their favorite movies, books, music, and sports.

7:26 FAMILY FAVORITES
A husband and wife try to guess each other's favorite kinds of movies, books, music, sports, and TV programs.
Host: Rich Young.

SBS-TV Backstage Bulletin Board

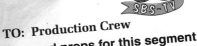

TO: Production Crew
Sets and props for this segment:

TV Studio
  cards   markers   microphone

TO: Cast Members
Key words in this segment:

book
movie
music
sport
TV program
What kind of . . . ?
favorite
like
adventure movie
cartoon
comedy
drama
science fiction
western
novel
poetry
short story
game show
news program

classical music
jazz
opera
popular music
rock music
baseball
football
golf
hockey
soccer
tennis
actor
actress
author
poet
performer

5:04 SBS-TV ON LOCATION

SCRIPT CHECK
Help the actors prepare their lines.

Movies
a. adventure movie   b. cartoon   c. comedy   d. drama
e. science fiction movie   f. western

1. __d__   2. ____   3. ____

4. ____   5. ____   6. ____

Books
a. novel   b. poetry   c. short story

7. ____   8. ____   9. ____

TV Programs
a. cartoon   b. comedy   c. drama   d. game show   e. news program

10. ____   11. ____   12. ____   13. ____   14. ____

Music

a. classical music   b. jazz   c. opera   d. popular music   e. rock music

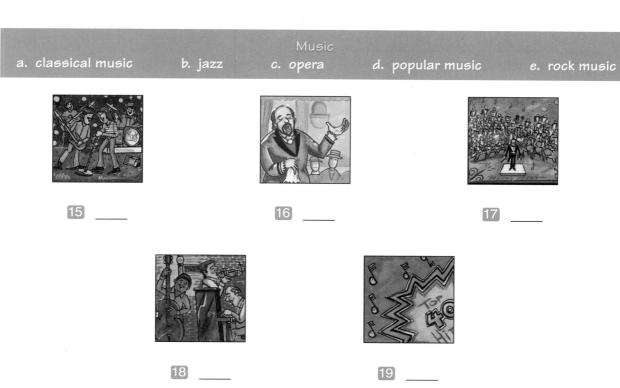

15 ____   16 ____   17 ____

18 ____   19 ____

Sports

a. baseball   b. football   c. golf   d. hockey   e. soccer   f. tennis

20 ____   21 ____   22 ____

23 ____   24 ____   25 ____

SEGMENT 15

CAN YOU PREDICT?

Before you watch the video scene, look at these people. What kind of movies do you think they like? Put a circle around each prediction. Then watch the interviews and see how many of your predictions are correct.

What kind of movies do you like?

1. a. (comedies)
   b. science fiction movies

2. a. adventure movies
   b. comedies

3. a. science fiction movies
   b. dramas

4. a. cartoons
   b. science fiction movies

5. a. westerns
   b. dramas

6. a. adventure movies
   b. cartoons

SOUND CHECK

Who is your favorite actor or actress?

1. a. (Goldie Hawn)
   b. Gilda Horn

2. a. Garrison Ward
   b. Harrison Ford

3. a. Marla Street
   b. Meryl Streep

4. a. Leonard Nimoy
   b. Bennet Nimoy

5. a. John Wang
   b. John Wayne

6. a. Mickey Mouse
   b. Mighty Mouse

SEGMENT 15

VIDEO EDITOR

You're editing the interviews with these people. Which words does each person say?

What kind of books do you like? Who is your favorite author or poet?

1. (a.) novels
   b. poetry
   c. Tom Clancy
   d. Tom Cranston

2. a. novels
   b. poetry
   c. Emily Nickerson
   d. Emily Dickinson

3. a. poetry
   b. short stories
   c. Edgar Allan Pope
   d. Edgar Allan Poe

CAN YOU PREDICT?

Before you watch the video scene, predict what these people are going to say. Then watch and check your answers.

What kind of TV programs do you like?

1. a. cartoons
   (b.) game shows

2. a. game shows
   b. news programs

3. a. comedies
   b. news programs

4. a. comedies
   b. dramas

5. a. cartoons
   b. comedies

WHAT'S MY LINE?

I like _____ 1.

My wife likes _____ 2.

SEGMENT 15

CAN YOU PREDICT?

Before you watch the video scene, predict what these people are going to say. Then watch and check your answers.

What kind of music do you like?

1. a. classical music
   (b.) rock music

2. a. jazz
   b. opera music

3. a. popular music
   b. rock music

4. a. classical music
   b. popular music

5. a. jazz
   b. opera music

SOUND CHECK

Who is your favorite performer?

1. a. Bryce Springstone
   (b.) Bruce Springsteen

2. a. Dave Brubeck
   b. Dave Brubaker

3. a. Margaret Dreisand
   b. Barbra Streisand

4. a. Zack Berlin
   b. Itzhak Perlman

5. a. Pavel Gotty
   b. Pavarotti

CAN YOU PREDICT?

Before you watch the video scene, predict what these people are going to say. Then watch and check your answers.

What's your favorite sport?

1. a. football
   b.) tennis

2. a. football
   b. baseball

3. a. baseball
   b. golf

4. a. baseball
   b. soccer

5. a. football
   b. golf

6. a. hockey
   b. tennis

CLOSE-UP

You're on Side by Side TV! Tell about YOUR favorites.

1. What kind of movies do you like? ....
2. Who is your favorite actor or actress? ....
3. What kind of books do you like? ....
4. Who is your favorite author? ....
5. What kind of TV programs do you like? ....
6. What kind of music do you like? ....
7. Who is your favorite performer? ....
8. What's your favorite sport? ....

INTERVIEW

Interview a friend. Ask about his or her favorites and write the answers below.

1. Movies: ....   3. TV Programs: ....
2. Books: ....   4. Music: ....
5. Sport: ....

SEGMENT 15

7:26 FAMILY FAVORITES

GUESTS AND HOST

1. Rich Young is ___.
   a. a contestant
   b.) the host

2. Dave and Donna Dawson live in ___.
   a. Denver
   b. Dallas

3. Dave is originally from ___.
   a. Denver
   b. Dallas

4. Donna is originally from ___.
   a. Des Moines
   b. Detroit

EDITING MIX-UP 1

The video editor made a mistake! Put the following lines in the correct order.

___ I know my wife. She likes comedies.

___ Comedies!

___ Congratulations, Dave and Donna. You're off to a good start. You have $100.

___ What is Donna's favorite kind of movie?

___ Okay, Donna and Dave. Time's up! Dave, what kind of movies does Donna like?

___ So Donna, what is your favorite kind of movie? Dave, what kind of movie is Donna's favorite? Write your answers down on the cards.

_1_ Do you have your cards and markers ready? I see you do. Then let's begin. It's time for our first question.

___ All right. Now let's see. Donna, what kind of movies do you like?

WHAT'S THE LINE?

1. a. What kind of books does Dave like?
   b. What kind of books Dave likes?

2. a. So Dave, what kind of books do you like?
   b. So Dave, what kind of books you do like?

3. a. Dave is liking novels.
   b. Dave likes novels.

4. a. He reads novels all the time.
   b. He reads novels all the time's up.

5. a. Dave, read you novels all the time?
   b. Dave, do you read novels all the time?

6. a. Yes, I do.
   b. Yes, I does.

SOUND CHECK 1

do	don't	I	you	like	jazz	kind	what
does	doesn't	she	your	likes	music	of	

RICH YOUNG: And the next question is: What kind of ___music___¹ does Donna _____²? So Donna, what _____³ of music do _____⁴ like? Dave, _____⁵ kind _____⁶ music does Donna _____⁷? Okay, Dave and Donna. Time's up! Let's see _____⁸ answers. Donna, what kind of music _____⁹ you _____¹⁰?

DONNA: I like _____¹¹.

RICH YOUNG: Dave, what kind of music _____¹² Donna _____¹³?

DAVE: She _____¹⁴ like jazz. She _____¹⁵ ROCK music.

DONNA: No, I _____¹⁶.

DAVE: Yes, you _____¹⁷.

RICH YOUNG: Dave, _____¹⁸ doesn't.

DAVE: Yes, she _____¹⁹.

DONNA: No, _____²⁰ don't.

SOUND CHECK 2

RICH YOUNG: What is Dave's favorite (sport/sports)¹? Dave, which sport (does/do)² (you/your)³ really (likes/like)⁴? Donna, Rich (which)⁵ sport (does/do)⁶ Dave (like/likes)⁷? Okay, Donna, let's see (your/you)⁸ answer. What (is/of)⁹ (Daves/Dave's)¹⁰ favorite sport?

DONNA: Baseball/Basketball¹¹.

RICH YOUNG: And Dave? (What/What's)¹² (your/you)¹³ favorite sport?

DAVE: Feetball/Football¹⁴.

DONNA: Football?! You (don't/doesn't)¹⁵ (like/likes)¹⁶ football!

DAVE: Of course I (does/do)¹⁷! I (LOVES/LOVE)¹⁸ football!

DONNA: No, you (don't/doesn't)¹⁹!

DAVE: Yes, I (does/do)²⁰! I (plays/play)²¹ football every (day/Saturday)²²!

EDITING MIX-UP 2

The video editor made another mistake! Put the following lines in the correct order.

____ Cartoons. Donna LOVES cartoons!

____ Yes, you do!

____ News programs!

____ No, I don't!

_1_ Dave, what kind of TV programs does Donna like?

____ Cartoons??!! I don't like cartoons!

____ News programs?! Hah! I watch news programs. YOU watch cartoons!

____ Donna, I'm almost afraid to ask. What's your answer?

SCRAMBLED WORDS

There are some problems with the sound track. Fix the scrambled words.

1. Our **aftervoi** sport is hockey.   favorite
2. Do you like **emocides**?
3. He reads **lovens** all the time.
4. Do you play **flog**?
5. My father writes **reptoy**.
6. I definitely like **niecces infotic**.
7. Are you listening to **scalliasc cuism**?
8. We really don't like **duverante viemos**.
9. Beverly Sills is a wonderful **frereprom**.
10. Our children like to watch **norcoast**.
11. I'm writing a **thors styro**.

ON CAMERA

You and a friend are on TV with Rich Young! Using the following as a guide, play "Student Favorites."

RICH YOUNG:	What kind of (movies/books/sports/music/TV programs) does .... like? (student's name)
STUDENT A:	He/She likes ....
RICH YOUNG:	.... , what kind of (movies/books/sports/music/TV programs) do you like? (student's name)
STUDENT B:	I like ....
RICH YOUNG:	(Congratulations!/I'm very sorry!) You now have .... dollars. Are you ready for the next question?
STUDENT A:	We're ready, Rich.
STUDENT B:	Yes, we're ready.

SEGMENT 15

SENTENCE CHALLENGE!

How many sentences can you make with the words below?

and	Donna	Dave	do	does	doesn't
don't	dramas	like	likes	but	?

1. ....
2. ....
3. ....
4. ....
5. ....
6. ....
7. ....
8. ....
9. ....
10. ....
11. ....
12. ....
13. ....
14. ....
15. ....
16. ....
17. ....
18. ....
19. ....
20. ....

AMERICAN CULTURE QUIZ

Do you know the names of American actors and actresses, musical performers, authors, and poets? Write the names below and compare your answers with your friends'.

Actor/Actress	Musical Performer	Author	Poet

SEGMENT 15 SCRIPT

5:04 SBS-TV ON LOCATION

INTERVIEWER: Do you go to the movies?
PERSON 1: Yes, I do.
INTERVIEWER: What kind of movies do you like?
PERSON 1: I like comedies.
INTERVIEWER: What kind of movies do you like?
PERSON 2: I like adventure movies.
PERSON 3: I like dramas.
PERSON 4: Hmm. I like science fiction movies.
PERSON 5: Westerns.
PERSON 6: I like cartoons.
INTERVIEWER: Who is your favorite actor or actress?
PERSON 1: Goldie Hawn. She's funny.
INTERVIEWER: Who is your favorite actor or actress?
PERSON 2: It's Harrison Ford. Definitely Harrison Ford.
PERSON 3: Meryl Streep.
PERSON 4: My favorite actor is Leonard Nimoy.
PERSON 5: John Wayne. I still like John Wayne. He's my favorite.
PERSON 6: You mean my favorite cartoon character? It's Mickey Mouse. Of course!

INTERVIEWER: Do you read a lot?
PERSON 7: Yes, I do.
INTERVIEWER: What kind of books do you like?
PERSON 7: I like novels.
INTERVIEWER: Who's your favorite author?
PERSON 7: Tom Clancy.
INTERVIEWER: What kind of books do you like?
PERSON 8: I like poetry.
INTERVIEWER: Poetry? That's nice. Who's your favorite poet?
PERSON 8: Emily Dickinson.
PERSON 9: I like short stories.
INTERVIEWER: Who's your favorite author?
PERSON 9: Edgar Allan Poe.

INTERVIEWER: Do you watch TV a lot?
PERSON 10: Yes, I do.
INTERVIEWER: What kind of TV programs do you like?
PERSON 10: I like game shows.
INTERVIEWER: What kind of TV programs do you like?
PERSON 11: I like news programs.
PERSON 1: I like comedies.
PERSON 5: Dramas.
PERSON 6: Cartoons. I like cartoons.

INTERVIEWER: What kind of music do you like?
PERSON 12: I like rock music.
INTERVIEWER: What kind of music do you like?
PERSON 7: I like jazz.
PERSON 13: I like popular music.
PERSON 8: Classical music.
PERSON 11: I like opera music.
INTERVIEWER: Who is your favorite performer?
PERSON 12: Bruce Springsteen.
PERSON 7: I like Dave Brubeck. He's great.
PERSON 13: My favorite performer? Let me see. I guess it's Barbra Streisand.
PERSON 8: Itzhak Perlman.
PERSON 11: Pavarotti is my favorite. He's terrific.

INTERVIEWER: Do you like sports?
PERSON 14: Yes, I do.
INTERVIEWER: What's your favorite sport?

PERSON 14:	Tennis.
INTERVIEWER:	What's your favorite sport?
PERSON 6:	I like football.
PERSON 5:	I like baseball.
PERSON 2:	Soccer.
PERSON 7:	My favorite sport? Hmm. It's golf.
PERSON 15:	I love hockey. Hockey is definitely my favorite sport.

7:26 FAMILY FAVORITES

ANNOUNCER:	And now it's time to play the world's favorite game show, *Family Favorites.* And here's the world's favorite game show host: Rich Young.
RICH YOUNG:	Thank you, ladies and gentlemen. Thank you. Thank you. And welcome to another edition of *Family Favorites,* the game show where husbands and wives find out how much they really know about each other. Let's meet our two contestants. Let's say "Hello" to Dave and Donna Dawson!
	Welcome, Dave and Donna. Tell us, Dave, where do you and Donna live?
DAVE:	We live in Denver.
RICH YOUNG:	Are you originally from Denver?
DAVE:	No, we aren't. I'm originally from Dallas, and Donna's originally from Des Moines.
DONNA:	No, Dave, that's not right. I'm not from Des Moines. I'm from Detroit.
DAVE:	Oh, that's right.
RICH YOUNG:	Well, wherever you're from, we're glad you're both here today to play *Family Favorites!* Now, here's how we play our game. I'll ask you both a question about something one of you likes. You'll both write your answers down on the cards in front of you, and we'll see if your answers match. For every match, we'll give you one hundred dollars. Now, are you ready to play our game?
DONNA:	Ready, Rich.
DAVE:	Yes. I'm ready.
RICH YOUNG:	Do you have your cards and markers ready? I see you do. Then let's begin. It's time for our first question. What is Donna's favorite kind of movie? So Donna, what is your favorite kind of movie? Dave, what kind of movie is Donna's favorite? Write your answers down on the cards. Okay, Donna and Dave. Time's up! Dave, what kind of movies does Donna like?
DAVE:	I know my wife. She likes comedies.
RICH YOUNG:	All right. Now let's see. Donna, what kind of movies do you like?
DONNA:	Comedies!
RICH YOUNG:	Congratulations, Dave and Donna. You're off to a good start. You have one hundred dollars. And now for our next question. What kind of books does Dave like? So Dave, what kind of books do you like? Donna, what kind of books does Dave like? All right, folks. Time's up! Donna, what kind of books does Dave like?
DONNA:	Dave likes novels. He reads novels all the time.
RICH YOUNG:	Dave, do you read novels all the time?
DAVE:	Yes, I do.
RICH YOUNG:	Dave and Donna, you now have two hundred dollars!

Are you ready for the next question?
DONNA: Ready, Rich.
DAVE: Ready.
RICH YOUNG: And the next question is: What kind of music does Donna like? So Donna, what kind of music do you like? Dave, what kind of music does Donna like? Okay, Dave and Donna. Time's up! Let's see your answers. Donna, what kind of music do you like?
DONNA: I like jazz.
RICH YOUNG: Dave, what kind of music does Donna like?
DAVE: She doesn't like jazz. She likes ROCK music.
DONNA: No, I don't.
DAVE: Yes, you do.
RICH YOUNG: Dave, she doesn't.
DAVE: Yes, she does.
DONNA: No, I don't.
RICH YOUNG: Well, Donna and Dave, that's okay. You still have two hundred dollars, and we still have two more questions. Are you ready for the next question?
DAVE: Ready.
DONNA: I'm ready.
RICH YOUNG: What is Dave's favorite sport? Dave, which sport do you really like? Donna, which sport does Dave like? Okay, Donna, let's see your answer. What is Dave's favorite sport?
DONNA: Baseball.
RICH YOUNG: And Dave? What's your favorite sport?
DAVE: Football.
DONNA: Football?! You don't like football!
DAVE: Of course I do! I LOVE football!
DONNA: No, you don't!
DAVE: Yes, I do! I play football every Saturday!
DONNA: No, you don't.
RICH YOUNG: Okay, Dave and Donna. It's time for our last question. Are you ready? Well, I guess they are. The last question is: What is Donna's favorite kind of TV program? So Donna, what kind of TV programs do you like? Dave, what kind of TV programs does Donna like? All right, folks. Time is up! Dave, what kind of TV programs does Donna like?
DAVE: Cartoons. Donna LOVES cartoons!
DONNA: Cartoons??!! I don't like cartoons!
RICH YOUNG: Donna, I'm almost afraid to ask. What's your answer?
DONNA: News programs!
DAVE: News programs?! Hah! I watch news programs. YOU watch cartoons!
DONNA: No, I don't!
DAVE: Yes, you do!
DONNA: No, I don't!
DAVE: Yes, you do!
DONNA: No, I don't!
DAVE: Rich, she does! She does! She really does!
DONNA: Rich, don't listen to him!
RICH YOUNG: Well, ladies and gentlemen, I'm afraid that's all the time there is. Remember, I don't care about your favorite movie, book, sport, or music . . . but I DO hope your favorite TV game show is *Family Favorites!* I'm your host, Rich Young, saying good-bye for now. See you next time.
ANNOUNCER: *Family Favorites* is a Side by Side Television Production.

SEGMENT 15 SUMMARY

GRAMMAR

Simple Present Tense

What kind of music	do	I we you they	like?
	does	he she it	

I We You They	like/don't like	jazz.
He She It	likes/doesn't like	

Yes/No Questions

Do	I we you they	like jazz?
Does	he she it	

Short Answers

Yes,	I we you they	do.
	he she it	does.

No,	I we you they	don't.
	he she it	doesn't.

FUNCTIONS

Asking for and Reporting Information

Do you *go to the movies*?
   Yes, I do.
   No, I don't.

Inquiring about Likes/Dislikes

What kind of *movies* do you like?
Which *sports* do you like?

Do you like *comedies*?

Who is your favorite *actor*?
What's your favorite *sport*?

Expressing Likes

I like *novels*.
I love *hockey*.

He's *great*.

Pavarotti is my favorite.

Expressing Dislikes

I don't like *cartoons*.

Expressing Surprise–Disbelief

Football?!

Expressing Certainty

Hockey is definitely *my favorite sport*.

Hesitating

Hmm.
Let me see.

SEGMENT 16

- Daily Activities
- Object Pronouns
- Adverbs of Frequency
- Simple Present Tense

"She always calls, I usually write, we sometimes dance, we never fight . . . because we're Side by Side."

PROGRAM LISTINGS

13:55 THERE'S THE PHONE!
Two friends tell about how often they talk to their friends and family.

14:47 SBS-TV ON LOCATION
People tell about when they watch TV.

15:12 SHE USUALLY STUDIES IN THE LIBRARY
Two college friends ask about another.

15:22 SBS-TV ON LOCATION
People tell about things they always, usually, sometimes, rarely, and never do.

16:33 I ALWAYS GET TO WORK ON TIME—GrammarRap
The GrammarRappers magically appear in an office and rap about punctuality.

SBS-TV Backstage Bulletin Board

TO: Production Crew
Sets and props for this segment:

Kitchen
 tables
 chairs
 glasses
 telephone

Office
 chairs
 clock
 desks
 office supplies

Dorm Room
 books
 chair
 desk
 lamp

TO: Cast Members
Key words in this segment:

 always
 usually
 sometimes
 rarely
 never

THERE'S THE PHONE!

SOUND CHECK

| them | me | us | you | it | her | him |

A. How often does your boyfriend call ____you____ ¹?

B. He calls _____ ² every night. How often do you talk to your brother in college?

A. I talk to _____ ³ every Sunday.

B. My sister is in college, too. I talk to _____ ⁴ every Saturday.

A. Do you speak to your grandparents very often?

B. Yes. I call _____ ⁵ every weekend.

A. My grandparents call _____ ⁶ every Friday night.

B. There's the phone.

A. I'll get _____ ⁷. Hello.

C. ....

A. Oh hi, Uncle George! It's my Uncle George! How are you? It's so good to hear from _____ ⁸! How's Aunt Sally?

C. ....

A. That's good. Please say "Hello" to _____ ⁹ from _____ ¹⁰.

MISSING LINES

What do you think Uncle George said on the telephone to his niece? Fill in Uncle George's lines above.

CLOSE-UP

How often do you speak to friends and family members on the telephone? Who are the people you talk to?

14:47 SBS-TV ON LOCATION

SOUND CHECK

When do you watch TV?

100%	90%	50%	10%	0%
always	usually	sometimes	rarely	never

1. a. usually
   b. **always** ✓
   c. morning
   d. **evening** ✓

2. a. always
   b. usually
   c. morning
   d. evening

3. a. sometimes
   b. always
   c. eight
   d. late

4. a. rarely
   b. nearly

5. a. ever
   b. never

Do you ever watch TV while you eat?

6. _always_   7. _____   8. _____

9. _____   10. _____

SEGMENT 16

15:12 SHE USUALLY STUDIES IN THE LIBRARY

SCENE CHECK

1. Carmen ~~isn't~~ (is) in her room right now.
2. She ~~never~~ rarely studies in her room.
3. She usually ~~always~~ studies in the library.

15:22 SBS-TV ON LOCATION

SOUND CHECK 1

Do you sing in the shower?

1. I ___never___ sing in the shower.
2. I _____ sing in the shower.
3. I _____ sing in the shower.

SOUND CHECK 2

Who washes the dishes in your household?

| wash | do | clean | cook |

I usually ___wash___¹ the dishes. Sometimes my wife ___washes___² the dishes, but usually it's me.

My husband and I always _____³ the dishes together. He rarely _____⁴ the dishes. I usually _____⁵ them, and he dries them. We're a good team!

 I never _____ ⁶ the dishes. My roommate always _____ ⁷ them. I _____ ⁸ all the cooking. It's a good arrangement. He _____ ⁹. I _____ ¹⁰.

SOUND CHECK 3

Do you snore when you sleep?

1. I ___rarely___ snore when I sleep.

2. I _____ snore when I sleep.

3. I _____ snore when I sleep.

4. I _____ snore when I sleep.

5. I _____ snore, but my husband _____ snores.

CLOSE-UP

Tell about YOUR household.

Who usually washes and dries the dishes? Who usually cooks?
Who usually cleans? Who usually does food shopping? How often?

16:33 I ALWAYS GET TO WORK ON TIME

FINISH THE RAP!

| always | never | rarely | sometimes | usually |

I _____always_____¹ get to work on time.
I'm _____² here by eight.
I _____³ get here early.
I _____⁴ get here late.
No, I _____⁵ get here late.

He _____⁶ gets to work on time.
He's _____⁷ here by eight.
He _____⁸ gets here early.
He _____⁹ gets here late.
   No, I _____¹⁰ get here late!
Right! He _____ gets here late!

WRITE YOUR OWN RAP!

Write a GrammarRap about yourself. Then write a GrammarRap about a friend.
Are you always on time for class or for work? Are you ever late?

GrammarRap about Me

I _____ get to _____ on time.
I'm _____ here by _____.
I _____ get here early.
I _____ get here late.
No, I _____ get here late!

GrammarRap about My Friend

He/She _____ gets to _____ on time.
He's/She's _____ here by _____.
He/She _____ gets here early.
He/She _____ gets here late.
   No, I _____ get here late!
Right! He/She _____ gets here late!

SEGMENT 16

SCRAMBLED WORDS

There are some problems with the sound track. Fix the scrambled words.

1. Jenny **wlaysa** studies in the evening.   always
2. Andy **lyerar** calls his aunt and uncle.
3. Bob **tsioemsme** watches TV after dinner.
4. Bill and Helen **alulyus** eat lunch at work.
5. Judy **venre** washes the dishes.

WHAT'S MY LINE?

1. (wash) I never ____wash____ the dishes. My husband usually _____ them.
2. (clean) How often do you _____ your apartment?
3. (study) My daughter usually _____ hard every night.
4. (watch) I rarely _____ TV. My roommate _____ TV all the time!
5. (call) I _____ my sister very often, and she _____ me a lot, too.
6. (get) All the people in my office _____ to work on time. Do you usually _____ to work on time?
7. (sing) My husband always _____ in the shower. My children _____ in the shower, too.
8. (fix) Who _____ things in your house when they're broken? I never _____ things around MY house.
9. (wash) My next-door neighbor _____ his car every weekend. I never _____ my car! How about you? Do you _____ your car very often?

TELL ABOUT YOURSELF!

Complete the following. Then have a friend complete it and compare your answers.

I always _____.   I rarely _____.

I usually _____.   I never _____.

I sometimes _____.

SEGMENT 16

SEGMENT 16 SCRIPT

13:55 THERE'S THE PHONE!

FRIEND 1: How often does your boyfriend call you?
FRIEND 2: He calls me every night. How often do you talk to your brother in college?
FRIEND 1: I talk to him every Sunday.
FRIEND 2: My sister is in college, too. I talk to her every Saturday.
FRIEND 1: Do you speak to your grandparents very often?
FRIEND 2: Yes. I call them every weekend.
FRIEND 1: My grandparents call us every Friday night.
FRIEND 2: There's the phone.
FRIEND 1: I'll get it. Hello. Oh hi, Uncle George!
(To her friend.)
It's my Uncle George!
(To Uncle on phone.)
How are you? It's so good to hear from you! How's Aunt Sally? That's good. Please say "Hello" to her from me.
ANNOUNCER: We're the telephone company. We bring people closer together.

14:47 SBS-TV ON LOCATION

INTERVIEWER: When do you watch TV?
PERSON 1: I always watch TV in the evening.
PERSON 2: I usually watch TV in the morning.
PERSON 3: I sometimes watch TV late at night.
PERSON 4: I rarely watch TV.
PERSON 5: I never watch TV.

INTERVIEWER: Do you ever watch TV while you eat?
PERSON 6: Always.
PERSON 7: Usually.
PERSON 8: Sometimes.
PERSON 9: Rarely.
PERSON 10: Never.

15:12 SHE USUALLY STUDIES IN THE LIBRARY

STUDENT 1: Does Carmen usually study in her room?
STUDENT 2: No. She rarely studies in her room. She usually studies in the library.

15:22 SBS-TV ON LOCATION

INTERVIEWER: Do you sing in the shower?
PERSON 1: No. I never sing in the shower.
PERSON 2: Sing in the shower? Well, sometimes.
PERSON 3: I always sing in the shower! I sing opera! La . . . la-la-la-la-la-la-la-la-la!

INTERVIEWER: Who washes the dishes in your household?
PERSON 4: I usually wash the dishes. Sometimes my wife washes the dishes, but usually it's me.
PERSON 5: My husband and I always do the dishes together. He rarely washes the dishes. I usually wash them and he dries them. We're a good team!
PERSON 6: I never wash the dishes. My roommate always washes them. I do all the cooking. It's a good arrangement. He cleans. I cook.

INTERVIEWER: Do you snore when you sleep?
PERSON 7: Rarely.
PERSON 8: Snore? No! Never!
PERSON 9: Sometimes, I guess.
PERSON 10: Yes. I usually snore.
PERSON 11: I never snore. But my husband always snores. Sometimes it doesn't bother me at all, and sometimes it drives me crazy!

16:33 I ALWAYS GET TO WORK ON TIME—GrammarRap

I always get to work on time.
I'm usually here by eight.
I sometimes get here early.
I never get here late.
No, I never get here late.

He always gets to work on time.
He's usually here by eight.
He sometimes gets here early.
He rarely gets here late.

No, I NEVER get here late!
Right! He NEVER gets here late!

SEGMENT 16 SUMMARY

GRAMMAR

Object Pronouns

He calls	me him her it us you them	every night.

Adverbs of Frequency

I	always usually sometimes rarely never	clean my apartment.

Simple Present Tense: *s* vs. *non-s* Endings

I We You They	eat. read. wash.

He She It	eat**s**. read**s**. wash**es**.	[s] [z] [ɪz]

FUNCTIONS

Asking for and Reporting Information

How often *does your boyfriend call you?*
　He calls *me every night.*

Who *washes the dishes in your household?*
　I usually *wash the dishes.*
　My brother sometimes *washes them.*

Does *Carmen* usually *study in her room?*
　No. *She* rarely *studies in her room.*
　Yes. *She* usually *studies in the library.*

How's *Aunt Sally?*

When *do you watch TV?*
　I always *watch TV in the morning.*

Greeting People

Hello.
Hi, *Uncle George.*

SEGMENT 17

- Describing People
- Have/Has

"Dad has brown eyes, Mom's are blue. He looks like me. She looks like you . . . Our family's Side by Side."

PROGRAM LISTINGS

17:19 DID YOU HEAR THAT?
Sound effects give clues about things people have.

19:10 MY SISTER AND I
Two sisters tell how they are different.

SBS-TV Backstage Bulletin Board

TO: Production Crew
Sets and props for this segment:

Kitchen
  counter
  pet food

Living Room
  sofa
  coffee table

Doctor's Office
  stethoscope
  examining table

Bedroom
  bed
  toys

Dining Room
  table
  chairs

TO: Cast Members
Key words in this segment:

cat
dog
mice
cockroach
neighbor
problem
heart
electric guitar

talent
motorcycle
car
eyes
hair
sister
different
look like

musical instrument
house
apartment
piano
TV
friend

17:19 DID YOU HEAR THAT?

WHAT DO THEY HAVE?

Circle the things that they have.

1. **a.** a pet ⓐ
   b. a dog
   c. a cat ⓒ

2. a. noisy neighbors
   b. nosy neighbors
   c. quiet neighbors

3. a. a problem with his heart
   b. a very unusual heartbeat
   c. a problem with his guitar

4. a. a new car
   b. a very old car
   c. a noisy car

5. a. a musical instrument
   b. an electric guitar
   c. musical talent

6. a. a bicycle
   b. a motorcycle
   c. a car

7. a. cockroaches
   b. mice
   c. a cat

CLOSE-UP

1 Do you have a dog?	Yes, I do.	No, I don't.
2 Do you have a cat?	Yes, I do.	No, I don't.
3 Do you have a new car?	Yes, I do.	No, I don't.
4 Do you have a motorcycle?	Yes, I do.	No, I don't.
5 Do you have noisy neighbors?	Yes, I do.	No, I don't.
6 Do you have a musical instrument?	Yes, I do.	No, I don't.
7 What other things do you have?		

SOUND EFFECTS MIX-UP

The sound editor needs help. Match the sound effects with the scenes below.

a. **Meow!**   b. **CRASH!**   c. **Thump Thump!**
d. Honk Honk!   e. **Bow Wow!**   f. squeak squeak!

1 _f_   2 ____   3 ____

4 ____   5 ____   6 ____

WHAT'S MY LINE?

I	Do he	Does she	have it	has we	they

1. ____Do____ you _____ a musical instrument?

   Yes, _____ _____ a guitar.

2. _____ Barbara _____ a sister?

   No, but _____ _____ three brothers.

3. _____ you and your wife _____ a car?

   Yes, _____ _____ an old car.

4. _____ your apartment _____ a dining room?

   No, but _____ _____ a very large kitchen.

5. _____ your neighbors _____ a pet?

   Yes, _____ _____ a very noisy dog.

6. _____ your son _____ problems at school?

   Yes, _____ _____ problems with English.

19:10 MY SISTER AND I

SOUND CHECK

| has | long | brown | apartment | color | friends | sisters | I'm |
| have | short | | two | bicycle | dog | guitar | don't |

SISTERS 1 & 2: My sister and I look very different.

SISTER 1: I have blue eyes and she _____1_____ _____2_____ eyes.

SISTER 2: I _____3_____ short hair and she _____4_____ _____5_____ hair.

SISTER 1: _____6_____ tall.

SISTER 2: And I'm _____7_____.

SISTERS 1 & 2: As you can see, I _____8_____ look like my sister. We look very different.

SISTER 1: I _____9_____ a house and she _____10_____ an .

SISTER 2: She _____12_____ a cat and I _____13_____ a _____14_____.

SISTERS 1 & 2: We both _____15_____ musical instruments.

SISTER 1: I _____16_____ a piano and she _____17_____ a _____18_____.

SISTER 2: She _____19_____ a car and I _____20_____ a _____21_____.

SISTER 1: I have a black-and-white TV and she _____22_____ a _____23_____ TV.

SISTER 2: She _____24_____ a lot of friends and I _____25_____ just one or _____26_____.

SISTER 1 & 2: As you can see, we're very different. But we're _____27_____, and we're _____28_____.

SEGMENT 17

WHICH SISTER?

1. She has (blue (brown)) eyes.
2. She has (blue   brown) eyes.
3. She has (short   long) hair.
4. She has (short   long) hair.
5. She's (tall   short).
6. She's (tall   short).
7. She has (a house   an apartment).
8. She has (a house   an apartment).
9. She has a (cat   dog).
10. She has a (cat   dog).
11. She has a (piano   guitar).
12. She has a (piano   guitar).
13. She has a (car   bicycle).
14. She has a (car   bicycle).
15. She has a (black-and-white   color) TV.
16. She has a (black-and-white   color) TV.
17. She has (a lot of   just one or two) friends.
18. She has (a lot of   just one or two) friends.

CLOSE-UP

You're on Side by Side TV! Tell about yourself. What color eyes do you have? Do you have short hair or long hair? Are you tall or short? Do you live in a house or an apartment? Do you have any pets? Do you have any musical instruments? Then tell about someone else—a friend, a family member, or even a famous person! How are you different?

TV CROSSWORD

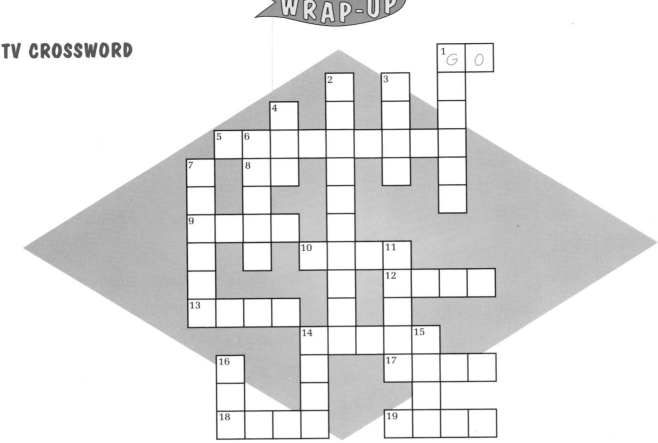

Across →

1. How often do you _____ to the movies?
5. We live in a nice _____.
8. I'm noisy, and my brother _____ quiet.
9. Do you have _____ hair?
10. My father and I both have brown _____.
12. Our children both have short _____.
13. Do your parents have noisy neighbors? Yes, _____ do.
14. Do I have a problem with my _____?
17. I _____ on the telephone a lot.
18. My apartment has _____!
19. My daughter has _____ eyes.

Down ↓

1. My brother has an electric _____.
2. Do you have a car or a _____?
3. _____ your son have a dog?
4. My sister _____ a very old bicycle.
6. The _____ is my favorite musical instrument.
7. Pavarotti has great musical _____.
11. My mother is tall, and I'm _____.
14. Mr. and Mrs. Smith _____ a new car.
15. Is she short or _____?
16. I write to _____ every week.

SEGMENT 17 SCRIPT

17:19 DID YOU HEAR THAT?

FRIEND 1: Do you have a dog?
FRIEND 2: No. I have a cat.

VISITOR: Do you have quiet neighbors?
WIFE: No. We have noisy neighbors.

PATIENT: Well, Doctor, do I have a problem with my heart?
DOCTOR: Yes. You have a VERY unusual heartbeat.
PATIENT: Oh, my!

BOY 1: Do your parents have a new car?
BOY 2: No. They have a very old car.

WIFE 1: Does your son have a musical instrument?
HUSBAND 2: Yes. He has an electric guitar.
WIFE 2: He has great musical talent. Don't you think so?
HUSBAND 1: Oh, yes. He does.
WIFE 1: Definitely!

FRIEND 1: Does your daughter have a car?
FRIEND 2: No. She has a motorcycle.

FRIEND 3: Does your apartment have cockroaches?
FRIEND 4: No. It has mice.

19:10 MY SISTER AND I

SISTERS 1 & 2: My sister and I look very different.
SISTER 1: I have blue eyes and she has brown eyes.
SISTER 2: I have short hair and she has long hair.
SISTER 1: I'm tall.
SISTER 2: And I'm short.
SISTERS 1 & 2: As you can see, I don't look like my sister. We look very different.
SISTER 1: I have a house and she has an apartment.
SISTER 2: She has a cat and I have a dog.
SISTERS 1 & 2: We both have musical instruments.
SISTER 1: I have a piano and she has a guitar.
SISTER 2: She has a car and I have a bicycle.
SISTER 1: I have a black-and-white TV and she has a color TV.
SISTER 2: She has a lot of friends and I have just one or two.
SISTERS 1 & 2: As you can see, we're very different. But we're sisters, and we're friends.

SEGMENT 17 SUMMARY

GRAMMAR

Have/Has

I / We / You / They	have	
He / She / It	has	brown eyes.

FUNCTIONS

Asking for and Reporting Information

Do you have *quiet neighbors?*

I have *a house.*
She has *an apartment.*

Describing

We *have noisy neighbors.*
They *have a very old car.*

My sister and I look very different.
I don't look like *my sister.*

We're very different.

I have *blue* eyes.
She has *brown* eyes.

I have *short* hair.
She has *long* hair.

I'm *tall.*

Expressing Certainty

Definitely!

Expressing Surprise–Disbelief

Oh, my!

Inquiring about Agreement

Don't you think so?

SEGMENT 18

- Feelings and Emotions
- Simple Present vs. Present Continuous
- Adjectives

"No one's angry, no one's sad, we're all happy and we're glad . . . to be Side by Side."

PROGRAM LISTINGS

20:13 I ALWAYS CRY WHEN I'M SAD
A man cries as he watches a very sad television program.

20:33 WE ALWAYS SHIVER WHEN WE'RE COLD
The heat isn't working in this apartment.

20:42 WHO'S THERE?
An unexpected visitor asks a lot of questions.

21:35 THE AUDITION
An actor acts out feelings and emotions at an audition.

22:51 SBS-TV ON LOCATION
People tell what they do when they're nervous, angry, and happy.

23:38 I SMILE WHEN I'M HAPPY—GrammarRap
The GrammarRappers rap about feelings.

SBS-TV Backstage Bulletin Board

TO: Production Crew
Sets and props for this segment:

Living Room
  chair
  rocking chair
  sofa
  television

Theater
  director's chair

TO: Cast Members
Key words for this segment:

bite	angry	nervous
cry	cold	sad
giggle	embarrassed	sick
perspire	happy	thirsty
shiver	hot	tired
shout	hungry	
smile		
whistle		
yawn		

20:13 I ALWAYS CRY WHEN I'M SAD

SOUND CHECK

| I | I'm | cry | crying |

Why are you _____crying_____¹?

_____ _____² because I'm sad.

_____³ always _____⁴ when

_____⁵ sad.

20:33 WE ALWAYS SHIVER WHEN WE'RE COLD

SOUND CHECK

| We | we're | shiver | shivering |

Why are you _____shivering_____¹?

_____ _____² because we're cold.

_____³ always _____⁴ when

_____⁵ cold.

TELL ME WHY!

| he | she | they |
| he's | she's | they're |

| blush | sing | dance |

___He's___¹ blushing _____⁵ singing _____¹⁰ dancing

because he's embarrassed. because _____⁶ happy. because _____¹¹ happy.

_____² always _____⁷ always _____¹² always

_____³ when _____⁸ when _____¹³ when

_____⁴ embarrassed. _____⁹ happy. _____¹⁴ happy.

20:42 WHO'S THERE?

SOUND CHECK

| I | I'm | ask | leave | shout | yawn |

Why are you ____yawning____¹?

Who's there?

It's me. You?

Yes. Why are you _____²?

I'm _____³ because _____⁴ tired. _____⁵ always _____⁶ when _____⁷ tired. And why are you _____⁸ me questions?

I always _____⁹ questions. It's my job.

Well, please _____¹⁰ right now! Go! Get out of here!

Why are you _____¹¹?

I'm _____¹² because _____¹³ angry. _____¹⁴ always _____¹⁵ when _____¹⁶ angry. Now please _____¹⁷! Go! Good-bye! And take your little word with you!

Okay. We're _____¹⁸. Sorry to bother you.

Hmm!

SEGMENT 18

21:35 THE AUDITION

WHICH CAPTION?

Help the video editor match the caption with the adjective.

| angry | cold | embarrassed | happy | hot | hungry |
| nervous | | sad | sick | thirsty | tired |

1. nervous
2. _____
3. _____
4. _____
5. _____
6. _____
7. _____
8. _____
9. _____
10. _____
11. _____

WHAT'S MY LINE?

| thirsty | cold | angry | happy | nervous | hot | hungry |

1. Please open the window. It's very ____hot____ in here.
2. I'm _____. Let's eat!
3. I'm shouting because I'm _____.
4. Close the window. It's _____!
5. Wendy is _____ today. It's her birthday.
6. I'm really _____. Is there any lemonade in the refrigerator?
7. William has a job interview tomorrow, and he's very _____.

22:51 SBS-TV ON LOCATION

SCRAMBLED SOUND TRACK

The sound track is all mixed up. Put the words in the correct order.

a. I'm . nervous , giggle I When   When I'm nervous, I giggle.
b. angry I . shout When I'm ,
c. I'm smile . When , happy I
d. nails I . When nervous bite I'm , my
e. get never angry . I
f. turns angry red . When my , I'm face
g. happy I'm When sing I . ,
h. When . nervous , I'm perspire I
i. happy When I whistle I'm . ,

MATCH THE LINES

Match the scrambled sound track above with the people in the interviews.

1. _d_   2. ____   3. ____

4. ____   5. ____   6. ____

7. ____   8. ____   9. ____

SEGMENT 18

23:38 I SMILE WHEN I'M HAPPY

FINISH THE RAP!

| blush | frown | shout | smile | embarrassed | mad |
| blushing | frowning | shouting | smiling | happy | sad |

I ____smile____¹ when I'm happy.
I _____² when I'm sad.
  I _____³ when I'm embarrassed.
And I _____⁴ when I'm mad.

Are you _____⁵?
  Yes. I'm happy.
Are you _____⁶?
  Yes. I'm sad.
  Are you _____⁷?
I'm embarrassed.
  Are you _____⁸?
Yes. I'm mad.

We smile when we're _____⁹.
We frown when we're _____¹⁰.
We blush when we're _____¹¹.
And we shout when we're _____¹².
We _____¹³ when we're happy.
We _____¹⁴ when we're sad.
We _____¹⁵ when we're embarrassed.
And we _____¹⁶ when we're mad.

WRITE YOUR OWN RAP!

I .... when I'm ....
I .... when I'm ....
I .... when I'm ....
And I .... when I'm ....

SEGMENT 18

WHY ARE THEY FEELING THIS WAY?

1. The students in my English class are happy today because ....

2. Betty is nervous today because ....

3. My neighbors are angry because ....

4. I'm very tired because ....

5. All my friends are sad today because ....

6. I'm embarrassed because ....

WHAT'S MY LINE?

1. When (I / I'm) nervous, I bite my nails.

2. Oh, really? I never (bite / biting) my nails.

3. I'm (smile / smiling) because I'm very happy.

4. I also (smile / smiling) when I'm happy.

5. When I'm angry, my face (turns / is turning) red.

6. Look! Your face (is turning / turns) red now!

7. (I relax / I'm relaxing) because I'm tired.

8. When I'm tired, I (sleeping / sleep).

9. (I / I'm) usually cry when I'm sad.

10. I never (crying / cry).

11. Why are you (giggle / giggling)?

12. (I / I'm) nervous.

13. Why are you (ask / asking) me questions?

14. I always (asking / ask) questions!

SEGMENT 18

SEGMENT 18 SCRIPT

20:13 I ALWAYS CRY WHEN I'M SAD

ANNOUNCER: Tune in tomorrow for another episode of *Our Children, Our Lives.*
INTERVIEWER: Why are you crying?
SAD MAN: I'm crying because I'm sad. I always cry when I'm sad.

20:33 WE ALWAYS SHIVER WHEN WE'RE COLD

INTERVIEWER: Why are you shivering?
WIFE: We're shivering because we're cold. We always shiver when we're cold.

20:42 WHO'S THERE?

INTERVIEWER: Why are you yawning?
TIRED LADY: Who's there?
INTERVIEWER: It's me.
TIRED LADY: You?
INTERVIEWER: Yes. Why are you yawning?

TIRED LADY: I'm yawning because I'm tired. I always yawn when I'm tired. And why are you asking me questions?
INTERVIEWER: I always ask questions. It's my job.
ANGRY LADY: Well, please leave right now! Go! Get out of here!
INTERVIEWER: Why are you shouting?
ANGRY LADY: I'm shouting because I'm angry. I always shout when I'm angry. Now please leave! Go! Good-bye! And take your little word with you!
INTERVIEWER: Okay. We're leaving. Sorry to bother you.
ANGRY LADY: Hmm!

21:35 THE AUDITION

DIRECTOR: All right. Who's next?
ACTOR: I am.
DIRECTOR: Oh, I remember you. Your name is Benny.
ACTOR: Lenny. Lenny Thomas.
DIRECTOR: All right, Lenny. I'm going to give you some adjectives . . .
ACTOR: More adjectives?
DIRECTOR: Yes . . . and you're going to act them out. Okay?
ACTOR: Yes. I'm ready.
DIRECTOR: Then let's begin. "Nervous." "Sad." "Happy." "Tired." "Sick." "Cold." "Hot." "Hungry." "Thirsty." "Angry." "Embarrassed." All right. Very good, Lenny. Thank you.
ACTOR: Is that all?
DIRECTOR: Yes, that's all. Thank you. Next!

22:51 SBS-TV ON LOCATION

INTERVIEWER: What do you do when you're nervous?
PERSON 1: I bite my nails.
PERSON 2: When I'm nervous? Let me see. I perspire.
PERSON 3: I giggle. I guess I'm nervous now.

INTERVIEWER: What do you do when you're angry?
PERSON 4: I shout.
PERSON 5: When I'm angry, my face turns red.
PERSON 6: Hmm. I don't know. I never get angry.

INTERVIEWER: What do you do when you're happy?
PERSON 7: I whistle.
PERSON 8: What do I do when I'm happy? That's easy. I smile.
PERSON 9: When I'm happy, I sing! Like this: la-la-la-la!

23:38 I SMILE WHEN I'M HAPPY—GrammarRap

I smile when I'm happy.
I frown when I'm sad.
  I blush when I'm embarrassed.
  And I shout when I'm mad.

Are you smiling?
  Yes. I'm happy.
Are you frowning?
  Yes. I'm sad.
  Are you blushing?
I'm embarrassed.
  Are you shouting?
Yes. I'm mad.

We smile when we're happy.
We frown when we're sad.
We blush when we're embarrassed.
And we shout when we're mad.

We smile when we're happy.
We frown when we're sad.
We blush when we're embarrassed.
And we shout when we're mad.

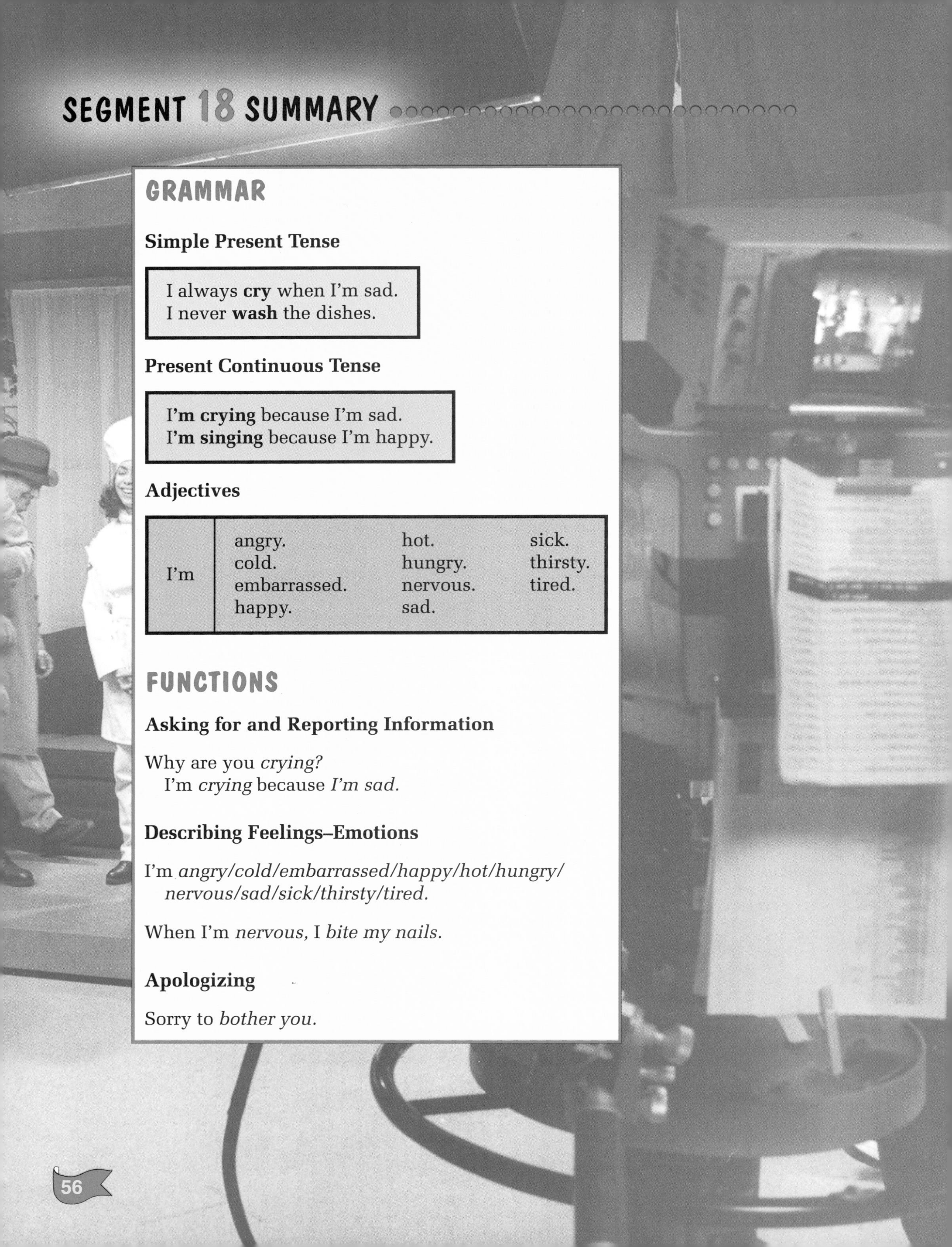

SEGMENT 18 SUMMARY

GRAMMAR

Simple Present Tense

> I always **cry** when I'm sad.
> I never **wash** the dishes.

Present Continuous Tense

> **I'm crying** because I'm sad.
> **I'm singing** because I'm happy.

Adjectives

I'm	angry. cold. embarrassed. happy.	hot. hungry. nervous. sad.	sick. thirsty. tired.

FUNCTIONS

Asking for and Reporting Information

Why are you *crying*?
  I'm *crying* because *I'm sad*.

Describing Feelings–Emotions

I'm *angry/cold/embarrassed/happy/hot/hungry/nervous/sad/sick/thirsty/tired*.

When I'm *nervous*, I *bite my nails*.

Apologizing

Sorry to *bother you*.

SEGMENT 19

- Describing Activities
- Simple Present vs. Present Continuous

"They're doing funny things today. They never do these things this way . . . They're Side by Side."

PROGRAM LISTINGS

24:38 I'M WASHING THE DISHES IN THE BATHTUB
Someone is very surprised when she sees what her friend is doing.

25:13 SPARKLE FLOOR CLEANER
One friend convinces another to switch from Ordinary Soap to Sparkle.

25:54 WHAT ARE THEY DOING?—GrammarRap
The GrammarRappers magically appear in an office, a kitchen, and a backyard.

SBS-TV Backstage Bulletin Board

TO: Production Crew
Sets and props for this segment:

Bathroom
  bathtub
  dishes
  dishwashing liquid
  rubber duck

Kitchen
  bucket
  mop
  floor cleaner

Office
  desk
  lamp
  clock

Kitchen
  stove
  pot
  spaghetti

Yard
  cat
  tub

TO: Cast Members
Key words in this segment:

bathtub
cat
dishes
floor
sink
soap
spaghetti

bathe
cook
shine
wash
work

broken
sorry
strange
late
always
never
usually

24:38 I'M WASHING THE DISHES IN THE BATHTUB

EDITING MIX-UP

The video editor made a mistake! Put the following lines in the correct order.

___ That's strange! Do you USUALLY wash the dishes in the bathtub?

___ Why are you doing THAT?!

_1_ What are you doing?!

___ Because my sink is broken.

___ I'm sorry to hear that.

___ No. I NEVER wash the dishes in the bathtub, but I'm washing the dishes in the bathtub TODAY.

___ I'm washing the dishes in the bathtub.

SOUND CHECK

A. What (are)¹ / do   you   do / doing ² ?!

B. I ³ / I'm   wash / washing ⁴ the dishes in the bathtub.

A. That's / What's ⁵ strange! Are / Do ⁶ you USUALLY wash / washing ⁷ the dishes in the bathtub?

B. No. I ⁸ / I'm   NEVER   wash / washing ⁹ the dishes in the bathtub,

but I ¹⁰ / I'm   wash / washing ¹¹ the dishes in the bathtub   TODAY / TUESDAY ¹² .

A. Why   are / do ¹³   you   do / doing ¹⁴   THAT?!

B. Because my sink   is / it's ¹⁵   broken.

A. I / I'm ¹⁶ sorry to   ear / hear ¹⁷ that.

25:13 SPARKLE FLOOR CLEANER

EDITING MIX-UP

The video editor made a mistake! Put the following lines in the correct order.

____ But you're using "Ordinary Soap!!"

____ Never! I use "Sparkle." Here, Tim. I have some with me. Let's try it on your floor right now.

_1_ Hi, Tim!

____ No more "Ordinary Soap" for me!

____ Wow! Look at this floor! It's shining!

____ Tim!! What are you doing?

____ Floors always shine with "Sparkle!"

____ Oh hi, Charles!

____ I'm washing my kitchen floor.

____ It always shines with "Sparkle!"

____ So? I always use "Ordinary Soap." Don't YOU use "Ordinary Soap" when you wash YOUR kitchen floor?

WHAT'S THE LINE?

brush	drink	feed	wash
brushing	drinking	feeding	washing

1. Why are you ___washing___ your hair with "Ordinary Shampoo?" I always _____ my hair with "Dazzle Shampoo."

2. Why are you _____ "Ordinary Coffee?" I never _____ "Ordinary Coffee." I _____ "Roaster's Choice."

3. But you're _____ your teeth with "Ordinary Toothpaste!" Why don't you _____ with "Smilodent?"

4. My dog is very important to me, so I always _____ him "Peppy Dog Food." How about you? What are you _____ your dog today?

SEGMENT 19

WRITE YOUR OWN COMMERCIAL!

Using the following as a guide, write a commercial for a cleaning product and practice it with a friend.

A. Hi, _____!
(friend's name)

B. Oh hi, _____!
(your name)

A. _____!! What are you doing?
(friend's name)

B. I'm _____ing my _____.

A. But you're using "_____!!"
(name of friend's cleaning product)

B. So? I always use "_____." Don't YOU use
(name of friend's cleaning product)

"_____" when you _____
(name of friend's cleaning product)

YOUR _____?

A. Never! I use "_____." Here, _____.
(name of your cleaning product) (friend's name)

I have some with me. Let's try it on your _____ right now.

B. Wow! Look at (this/these) _____! (It's/They're) shining!

A. (It/They) ALWAYS (shines/shine) with "_____!"
(name of your cleaning product)

B. No more "_____" for me!
(name of friend's cleaning product)

A. _____s ALWAYS shine with "_____!"
(name of your cleaning product)

WHAT ARE THEY DOING?

FINISH THE RAP!

always	bathing	cooks	cooking	doing	he	he's	his
it's	late	on	that	what's	why's	works	working

What's Fran ___doing___¹?
  She's _____² late.
Working _____³?
Why's she _____⁴ that?
  It's Monday. She always _____⁵ late on Monday.

_____⁶ Bob doing?
_____⁷ cooking spaghetti.
_____⁸ spaghetti?
Why's _____⁹ doing _____¹⁰?
  It's Wednesday. _____¹¹ always _____¹² spaghetti _____¹³ Wednesday.

What's Gary _____¹⁴?
  _____¹⁵ bathing _____¹⁶ cat.
  _____¹⁷ _____¹⁸ cat?
  _____¹⁹ he doing that?
  _____²⁰ Friday. He _____²¹ bathes _____²² cat on Friday.

WRITE YOUR OWN RAP!

What are you doing?
  I'm ....
  ....? Why are you doing that?
  It's .... I always ....
  on ....

WRAP-UP

WRONG LINE
Cross out the mistakes.

1. Mary always ~~study~~ / studying / studies English ~~on~~ / at / ~~in~~ Sunday.

2. She / She's / ~~Her~~ bathing the ~~dog~~ / cat / ~~dishes~~.

3. They're cleaning their house ~~today~~ / ~~usually~~ / right now.

4. ~~Are~~ / Do / Does you usually ~~eat~~ / ~~eats~~ / eating breakfast?

5. Our children never ~~watch~~ / watches / ~~watching~~ TV on ~~Monday~~ / Wednesday / ~~today~~.

6. Floors always ~~shine~~ / shine / ~~shining~~ with "Sparkle."

7. But ~~you~~ / you're / ~~your~~ using "Ordinary Soap!"

8. Why's ~~he~~ / they / ~~you~~ doing that?

SCRAMBLED SOUND TRACK
The sound track is all mixed up. Put the words in the correct order.

1. apartment | always | cleans | . | on | He | his | Friday

   He always cleans his apartment on Friday.

2. in | the | usually | you | Do | dishes | the | ? | wash | bathtub

   _____

3. to | her | walking | . | bicycle | school | is | because | She's | broken

   _____

4. I | working | late | never | but | I'm | today | late | work | , | .

   _____

5. spaghetti | ? | cook | Does | on | usually | he | Wednesday

   _____

6. "Ordinary Soap?" | Why | washing | dishes | are | with | your | you

   _____

SEGMENT 19 SCRIPT

24:38 I'M WASHING THE DISHES IN THE BATHTUB

FRIEND 1: What are you doing?!
FRIEND 2: I'm washing the dishes in the bathtub.
FRIEND 1: That's strange! Do you USUALLY wash the dishes in the bathtub?
FRIEND 2: No. I NEVER wash the dishes in the bathtub, but I'm washing the dishes in the bathtub TODAY.
FRIEND 1: Why are you doing THAT?!
FRIEND 2: Because my sink is broken.
FRIEND 1: I'm sorry to hear that.

25:13 SPARKLE FLOOR CLEANER

CHARLES: Hi, Tim!
TIM: Oh hi, Charles!
CHARLES: Tim!! What are you doing?
TIM: I'm washing my kitchen floor.
CHARLES: But you're using "Ordinary Soap!!"
TIM: So? I always use "Ordinary Soap." Don't YOU use "Ordinary Soap" when you wash YOUR kitchen floor?
CHARLES: Never! I use "Sparkle." Here, Tim. I have some with me. Let's try it on your floor right now.

(Charles pours some "Sparkle" on the floor and takes the mop from Tim's hands.)

TIM: Wow! Look at this floor! It's shining!
CHARLES: It always shines with "Sparkle!"
TIM: No more "Ordinary Soap" for me!
ANNOUNCER: Floors always shine with "Sparkle!"

25:54 WHAT ARE THEY DOING?— GrammarRap

What's Fran doing?
  She's working late.
Working late?
Why's she doing that?
  It's Monday.
  She always works late on Monday.

What's Bob doing?
  He's cooking spaghetti.
Cooking spaghetti?
Why's he doing that?
  It's Wednesday.
  He always cooks spaghetti on Wednesday.

What's Gary doing?
  He's bathing his cat.
Bathing his cat?
Why's he doing that?
  It's Friday.
  He always bathes his cat on Friday.

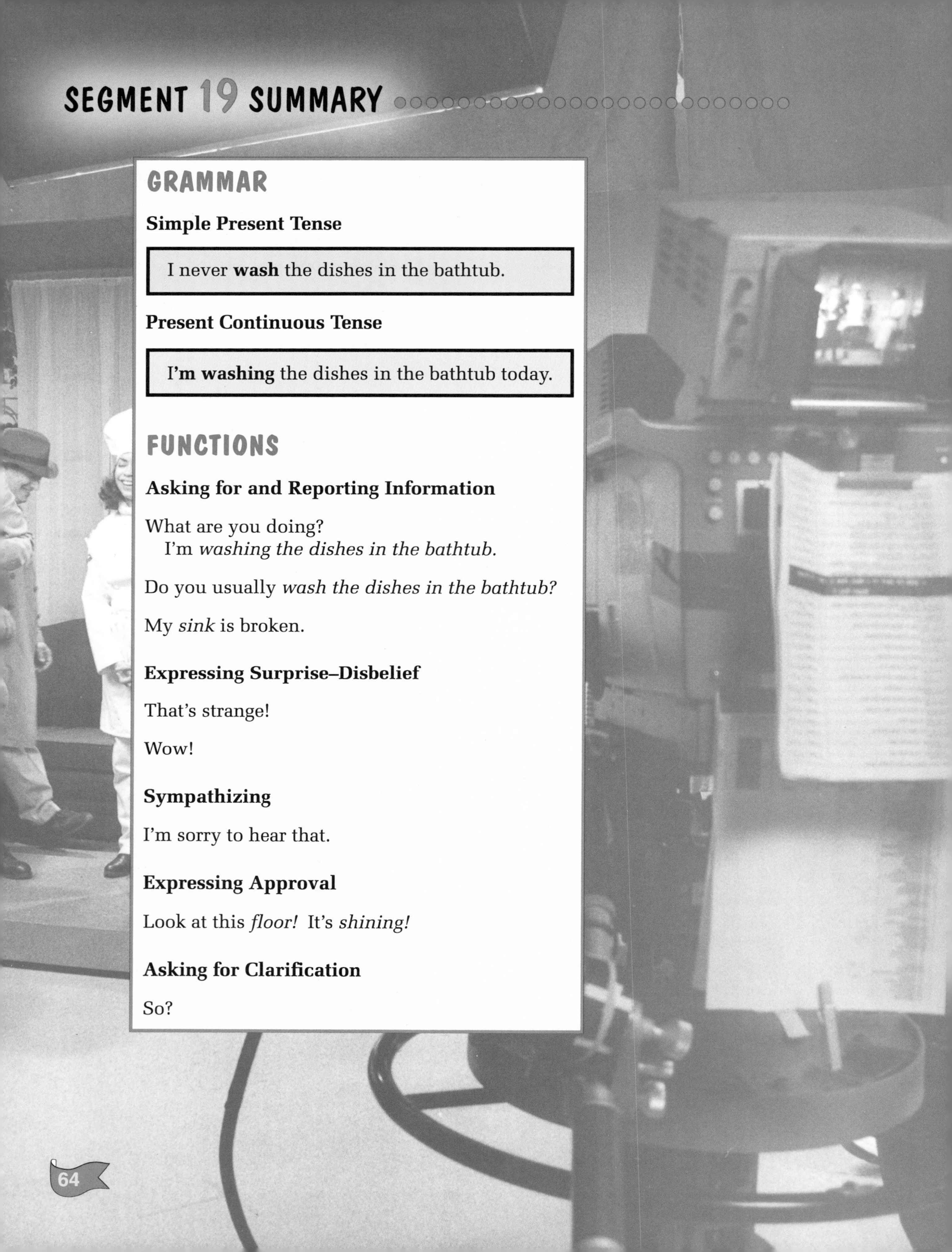

SEGMENT 19 SUMMARY

GRAMMAR

Simple Present Tense

> I never **wash** the dishes in the bathtub.

Present Continuous Tense

> I**'m washing** the dishes in the bathtub today.

FUNCTIONS

Asking for and Reporting Information

What are you doing?
   I'm *washing the dishes in the bathtub.*

Do you usually *wash the dishes in the bathtub?*

My *sink* is broken.

Expressing Surprise–Disbelief

That's strange!

Wow!

Sympathizing

I'm sorry to hear that.

Expressing Approval

Look at this *floor!* It's *shining!*

Asking for Clarification

So?

SEGMENT 20

- Expressing Ability
- Occupations
- Can

"They can dance and they can sing. They can do most anything . . . together Side by Side."

PROGRAM LISTINGS

27:20 SBS-TV ON LOCATION
People tell about things they can and can't do.

28:25 CAN YOU?
Two diplomats meet in front of the United Nations and discuss languages.

28:32 SBS-TV ON LOCATION
People tell about languages they can speak.

28:59 OF COURSE THEY CAN
A customer is concerned about a mechanic's ability to fix his car.

29:25 SO YOU'RE LOOKING FOR A JOB
A job seeker goes to the Ace Employment Service.

SBS-TV Backstage Bulletin Board

TO: Production Crew
Sets and props for this segment:

United Nations
  sign

Auto Shop
  car
  tools

Employment Agency
  chairs
  computer
  desk

TO: Cast Members
Key words in this segment:

actor
baker
bus driver
chef
dancer
mechanic
painter
secretary
teacher
truck driver
writer

Chinese
Spanish
Japanese
Korean
Portuguese

sing
play
speak
fix
bake

drive
cook
type
teach
paint
dance
write
act

27:20 **SBS-TV ON LOCATION**

SOUND CHECK

Can you sing?

`can    can't`

1. Yes, I ___can___.

2. No, I _____.

   You see, I _____ sing.

Can you play the violin?

3. Yes, I _____.

4. No, I _____. I'm sorry.

   I just _____ play the violin.

WHAT CAN THEY DO?

1. Can she ski?
   - a. Yes, she can. ✓
   - b. No, she can't.

2. Can he drive?
   - a. Yes, he can.
   - b. No, he can't.

3. Can they dance?
   - a. Yes, they can.
   - b. No, they can't.

4. Can he type?
   - a. Yes, he can.
   - b. No, he can't.

5. Can she swim?
   - a. Yes, she can.
   - b. No, she can't.

6. Can they cook?
   - a. Yes, they can.
   - b. No, they can't.

SEGMENT 20

CAN YOU?

SOUND CHECK

A. ____Can____¹ you speak Hungarian?

B. No, I _____². But I _____³ speak Romanian.

SBS-TV ON LOCATION

INFORMATION CHECK

Circle the statements that are true about each person.

1
a. He can't speak English.
b.) He can speak English.
c.) He can speak Chinese.

2
a. He can't speak Chinese.
b. He can't speak Spanish.
c. He can speak Spanish.

3
a. She can't speak Japanese.
b. She can speak Japanese.
c. She can't speak Spanish.

4
a. He can't speak Korean.
b. He can speak Japanese.
c. He can't speak Japanese.

5
a. She can't speak Korean.
b. She can't speak Portuguese.
c. She can speak Portuguese.

6
a. She can speak today.
b. She can't speak today.
c. She can speak English.

CLOSE-UP

What languages can YOU speak?

28:59 OF COURSE THEY CAN

SCRAMBLED SOUND TRACK

The sound track is all mixed up. Put the words in the correct order.

fix	course	a	.
cars	every	?	.
Of	Jack	He	
Can	mechanic		
He's	can	cars	
he	fixes	day	!

A. Can Jack fix cars?

B. _____

_____

_____

_____

_____

SCRIPT CHECK

Help the cast rehearse important words in this segment.

| baker | chef | painter | teacher | |
| bus driver | dancer | secretary | truck driver | writer |

1. Mario bakes very well. He's a ___baker___.

2. Paul drives a truck. He's a _____.

3. Linda teaches. She's a _____.

4. Alberto can cook. He's a _____.

5. Sam can paint. He's a _____.

6. Julie writes very well. She's a _____.

7. Frank drives a bus. He's a _____.

8. Gloria dances very well. She's a _____.

9. Irene can type. She's a _____.

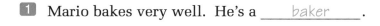

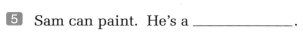
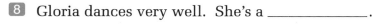

SEGMENT 20

29:25 SO YOU'RE LOOKING FOR A JOB

SCENE CHECK

What things **can** this job applicant do? What **can't** he do?

1. He (**can**) can't) fix cars.
2. He (can   can't) bake bread and cakes.
3. He (can   can't) bake chocolate chip cookies.
4. He (can   can't) drive a bus.
5. He (can   can't) drive a car.
6. He (can   can't) cook.
7. He (can   can't) type.
8. He (can   can't) teach.
9. He (can   can't) paint.
10. He (can   can't) drive a truck.
11. He (can   can't) dance.
12. He (can   can't) write.
13. He (can   can't) act.

EDITING MIX-UP

The video editor made a mistake! Put each pair of lines in the correct order.

1. __2__ I can fix cars.
   __1__ Tell me, what can you do?

2. ____ Yes, that's right.
   ____ Oh. You're a mechanic!

3. ____ Can you do anything else?
   ____ Well . . . I can do lots of things, I guess.

4. ____ No, I can't.
   ____ Can you bake bread and cakes?

5. ____ I can drive a car, but I can't drive a bus.
   ____ That's too bad.

6. ____ Oh, no! I can't cook at all!
   ____ Can you cook?

7. ____ I can't drive a truck.
   ____ This company is looking for a truck driver.

8. ____ Thank you.
   ____ Believe me, you can act. You're a terrific actor.

SEGMENT 20

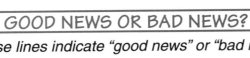

GOOD NEWS OR BAD NEWS?

Do these lines indicate "good news" or "bad news"?

1	"I can fix cars."	**Good news**	Bad news
2	"I'm afraid we don't have any jobs for mechanics right now."	Good news	Bad news
3	"I can do lots of things, I guess."	Good news	Bad news
4	"I'm sure we can find a job that's right for you."	Good news	Bad news
5	"I'm afraid I can't."	Good news	Bad news
6	"Oh, no! I can't cook at all!"	Good news	Bad news
7	"Here's a job for a secretary."	Good news	Bad news
8	"Sorry. I can't."	Good news	Bad news
9	"No, definitely not. I can't paint."	Good news	Bad news
10	"You can act!"	Good news	Bad news
11	"I think you can get the job."	Good news	Bad news

TO BE OR NOT TO BE AN ACTOR!

Look at the pictures and perform the scene . . . very dramatically!

CAN THEY OR CAN'T THEY?

1. Ben ___can___ type. He's a very good ___secretary___.
2. I ___can't___ write very well at all. I'm not a very good writer.
3. Olivia _____ bake bread and cakes. She's an excellent _____.
4. I'm not a very good _____. I _____ teach anything!
5. Jim is an excellent _____. He _____ fix any problem with a car.
6. I'm a _____. I _____ drive big trucks and small trucks.
7. Pierre is a _____ at a restaurant. He _____ cook delicious meals.
8. I'm a very bad _____. I _____ paint at all!

SURVEY

Which of these things can you do? Answer "Yes, I can" or "No, I can't."

1. Can you drive?
2. Can you cook?
3. Can you sing?
4. Can you dance?
5. Can you swim?
6. Can you ski?
7. Can you skate?
8. Can you draw?
9. Can you play the guitar?
10. Can you play basketball?
11. Can you ride a bicycle?
12. Can you ride a horse?
13. Can you type?
14. Can you use a computer?

INTERVIEW

Make up five questions and then interview two friends. What can your friends do?

	Friend 1	Friend 2
1. Can you _____?		
2. Can you _____?		
3. Can you _____?		
4. Can you _____?		
5. Can you _____?		

SEGMENT 20

SEGMENT 20 SCRIPT

27:20 SBS-TV ON LOCATION

INTERVIEWER: Can you sing?
PERSON 1: Yes, I can.
(Singing.)
People, people who need people, are the luckiest people in the world.
(Asking Interviewer.)
How's that?
INTERVIEWER: Very nice.

INTERVIEWER: Can you sing?
PERSON 2: No, I can't.
(Singing.)
Feelings . . . nothing more than feelings . . .
(To Interviewer.)
You see? I can't sing. Sorry.
INTERVIEWER: That's okay.

INTERVIEWER: Can you play the violin?
PERSON 3: Yes, I can.
(Person 3 plays the violin.)
INTERVIEWER: That's beautiful!
PERSON 3: Thank you.

INTERVIEWER: Can you play the violin?
PERSON 4: No, I can't.
(Person 4 plays the violin.)
PERSON 4: I'm sorry. I just can't play the violin.
INTERVIEWER: That's all right.

28:25 CAN YOU?

DIPLOMAT 1: Can you speak Hungarian?
DIPLOMAT 2: No, I can't. But I can speak Romanian.

28:32 SBS-TV ON LOCATION

INTERVIEWER: What languages can you speak?
PERSON 1: I can speak English and Chinese.
PERSON 2: I can't speak Chinese, but I can speak Spanish.
PERSON 3: I can't speak Spanish, but I can speak Japanese.
PERSON 4: Japanese? No. I can't speak Japanese, but I CAN speak Korean.
PERSON 5: No. I'm afraid I can't speak Korean, but I can speak Portuguese.
PERSON 6: I can't speak anything today. I have laryngitis. Sorry.

28:59 OF COURSE THEY CAN

CUSTOMER: Can Jack fix cars?
MANAGER: Of course he can. He fixes cars every day. He's a mechanic!

29:25 SO YOU'RE LOOKING FOR A JOB

AGENCY OWNER: So you're looking for a job.
JOB SEEKER: Yes, I am.
AGENCY OWNER: Well, tell me, what can you do?
JOB SEEKER: I can fix cars.
AGENCY OWNER: Oh. You're a mechanic!
JOB SEEKER: Yes, that's right.
AGENCY OWNER: Well, let's see if we can find a job for a mechanic. Hmm . . . I'm afraid we don't have any jobs for mechanics right now. Can you do anything else?
JOB SEEKER: Hmm. Well . . . I can do lots of things, I guess.
AGENCY OWER: Such as?
JOB SEEKER: Well, I . . . I can . . . uh . . . I can . . . Hmm.
AGENCY OWNER: Tell you what. Let's check and see what kinds of jobs we have right now. I'm sure we can find a job that's right for you.
JOB SEEKER: Thank you.
AGENCY OWNER: Let me see. Oh, here we are. Betty's Bakery is looking for a baker. Can you bake bread and cakes?
JOB SEEKER: No, I can't. I can bake chocolate chip cookies, but I can't bake bread or cakes.
AGENCY OWNER: All right. Let's see what else we have. Ah! The Ajax Bus Company is looking for a bus driver. Can you drive a bus?
JOB SEEKER: I'm afraid I can't. I can drive a car, but I can't drive a bus.
AGENCY OWNER: That's too bad. Let's keep looking here. Hmm. Here's one. The Renaissance Restaurant is looking for a chef. Can you cook?
JOB SEEKER: Oh, no! I can't cook at all!
AGENCY OWNER: Well, here's a job for a secretary. Can you type?
JOB SEEKER: No, I can't.
AGENCY OWNER: Hmm. This school is looking for a teacher. Can you teach?
JOB SEEKER: Sorry. I can't.
AGENCY OWNER: Here's one for a painter.
JOB SEEKER: No, definitely not. I can't paint.
AGENCY OWNER: This company is looking for a truck driver.
JOB SEEKER: I can't drive a truck.
AGENCY OWNER: This place needs a dancer.
JOB SEEKER: I can't dance.
AGENCY OWNER: Here's one for a writer.
JOB SEEKER: I can't write. This is terrible! I can't believe this! I can't bake, I can't drive a bus, I can't cook, I can't type, I can't teach, I can't paint, I can't drive a truck, I can't dance, I can't write. I can't do ANYTHING!! Woe is me! What can I do?! What can I do? What can I do?
AGENCY OWNER: You can act!
JOB SEEKER: I can what?
AGENCY OWNER: You can act! And I have a job for you! Here! The Stagelight Theater Company is looking for an actor.
JOB SEEKER: Me? An actor?
AGENCY OWNER: Yes. You're an excellent actor. Go down to the Stagelight Theater right now. I think you can get the job.
JOB SEEKER: Do you really think so?
AGENCY OWNER: Believe me, you can act. You're a terrific actor.
JOB SEEKER: Thank you. Thank you so much.

SEGMENT 20 SUMMARY

GRAMMAR

Can

Can	I / he / she / it / we / you / they	sing?

I / He / She / It / We / You / They	can / can't	dance.

Yes,	I / he / she / it / we / you / they	can.

No,	I / he / she / it / we / you / they	can't.

FUNCTIONS

Inquiring about Ability

Can you *speak Hungarian?*

Expressing Ability

Yes, I can.
I can *speak Japanese.*

Of course *he* can.

Expressing Inability

I can't *type.*
No, I can't.

Asking for and Reporting Information

He *fixes cars every day.*
He's a *mechanic.*

Tell me, _____.

Inquiring about Want–Desire

What kind of *job* are you looking for?

Complimenting

Very nice!
That's beautiful!
You're a terrific *actor.*

Hesitating

Hmm.
Well, . . .

Expressing Regret

That's too bad.

Sorry.
I'm sorry.

SEGMENT 21

- Obligations
- Invitations
- Have to
- Can

"I'm sorry I can't go with you. I have some things I have to do . . . Side by Side."

PROGRAM LISTINGS

33:00 THEY CAN'T GO TO HERBERT'S PARTY
Nobody can go to Herbert's party.

34:34 I'M SORRY. I CAN'T.
Steve, an office worker, refuses several invitations.

37:47 WE CAN'T TALK NOW!—GrammarRap
The GrammarRappers meet on a busy city sidewalk, but they're in a rush and don't have time to stop and talk.

SBS-TV Backstage Bulletin Board

TO: Production Crew
Sets and props for this segment:

Living Room
sofa
coffee table
refreshments
balloons

Office
desks
desk calendar
telephone
computer

Street
buses
cars
briefcases

TO: Cast Members
Key words in this segment:

do laundry
go bowling
go sailing
go shopping
catch a plane
catch a train
chat
make a call
stop

ballgame
dentist
doctor
party
tickets
depressed
sorry
tomorrow evening
tonight

33:00 THEY CAN'T GO TO HERBERT'S PARTY

SOUND CHECK

I	you	can	clean	doctor
I'm	your	can't	do	party
me	they	has	go	to
my	their	have	work	

Herbert is depressed. He's having a party today, but his friends <u>can't go</u>¹ to his party. They're all busy.

A. _____² you go to Herbert's party?

B. No, _____ _____³. I have to _____⁴.

A. Can you _____⁵ to Herbert's party?

B. No, we can't. We _____ _____ _____⁶ our house.

A. Well, _____ _____⁷ children _____⁸ to Herbert's party?

B. No, _____ _____⁹. They have to _____¹⁰ _____¹¹ homework.

A. Can you go to Herbert's _____¹²?

B. Yes, I _____¹³.

C. No, _____¹⁴ can't. You _____ _____ _____¹⁵ to the dentist. She _____¹⁶. She _____ _____ _____¹⁷ to the dentist.

B. That's right. I _____¹⁸.

A. Can Michael _____ _____ 19 Herbert's party?

B. No, he can't. He _____ _____ 20 go _____ 21 the _____ 22.

A. Herbert?

B. Yes?

A. _____ 23 sorry, Herbert. Your friends _____ 24 come to your party. They all have things they _____ _____ _____ 25.

B. Oh. Well, _____ _____ 26 come to _____ 27 party?

A. _____ 28? Oh, no. I'm afraid _____ _____ 29. I _____ _____ 30 stay here and _____ 31. Sorry.

B. That's okay. _____ 32 understand.

ON CAMERA

Herbert is inviting you and your family to his party, but you're all busy.

HERBERT: Can you come to my party?

YOU: No, I can't. I have to ....

HERBERT: Can your .... come to my party?

YOU: I'm sorry. (He/She/They) can't. (He/She/They) (has to/have to) ....

HERBERT: Well, can your .... come to my party?

YOU: I'm afraid (he/she/they) can't. (He/She/They) (has to/have to) ....

HERBERT: That's okay. I understand.

SEGMENT 21

34:34 I'M SORRY. I CAN'T.

INFORMATION CHECK

Watch the scene and then put a check (✔) next to the things Steve HAS TO do.

____ work late	____ go to a ballgame
____ do laundry	____ go to a party
____ go bowling	____ have dinner with Julie
____ go sailing	____ have dinner with Mom and Dad
✔ go shopping	____ clean his apartment

EDITING MIX-UP 1

The video editor made a mistake! Put Steve's conversation with Bob in the correct order.

____ Tonight? Let me see. Gee, no, I can't. I have to go shopping.

____ Fine, Bob.

____ Oh. That's too bad.

_1_ Oh hi, Steve! How are you today?

____ But thanks for asking me.

____ Hey, Steve, can you go to the ballgame with me tonight? I have two tickets.

____ Sure.

EDITING MIX-UP 2

Put Steve's conversation with Patty in the correct order.

____ Sorry you can't come.

____ Great. Listen, Steve, my roommate and I are having a party tomorrow evening. Can you come?

____ See you later.

____ Oh hi, Patty! How are you doing?

____ Me, too.

_1_ Hi, Steve!

____ Tomorrow evening? Let me check. Hmm. Looks like I can't. I have to do my laundry. But thanks for inviting me.

EDITING MIX-UP 3

Put Steve's conversation with Alan in the correct order.

____ This Saturday? Hmm. I don't know. Nope. I'm afraid not. I have to clean my apartment this Saturday.

__1__ Hi, Alan. How are you doing today?

____ See you later, Alan.

____ Yes, I do. I really have to clean it this Saturday.

____ Do you have to clean it on Saturday?

____ Well, all right, Steve. See you later.

____ Pretty good. Listen, Steve, can you go sailing this Saturday?

WHAT'S JULIE SAYING?

Listen to Steve's conversation with Julie. Can you guess what she's saying? Circle the best answer.

1 Hi, Julie. This is Steve.
  a. Hi, Steve. I'm calling from Miami.
  b. Hi, Steve. What are you doing?
  c.) Hi, Steve. How are you doing?

2 Listen, Julie, can you have dinner with me on Sunday?
  a. On Sunday? Let me see. Of course, I can.
  b. On Sunday? Gee, I'm sorry. I can't.
  c. On Sunday? I see.

3 You can't?
  a. I'm afraid not. I have to fix my TV.
  b. I'm sorry. I can't. I have to eat dinner.
  c. Me, too.

4 You have to WHAT?
  a. Because it's broken.
  b. I have to fix my TV.
  c. I have a color TV.

5 Do you have to do that on Sunday?
  a. Yes, I do.
  b. Yes, I can.
  c. Yes, I have.

6 I see. Well, that's too bad. Maybe some other time.
  a. I'm sorry to hear that.
  b. Thanks for asking me.
  c. See you on Sunday.

FRED'S MISSING LINES

What do you think Fred is saying to Steve? Fill in Fred's "missing lines" and then practice the conversation with a friend.

STEVE: Hi, Fred. This is Steve.

FRED: ....

STEVE: Great. Great. Listen, Fred, can you go bowling with me on Sunday?

FRED: ....

STEVE: That's too bad. I'm sorry you can't. And you have to do that on Sunday?

FRED: ....

STEVE: Of course I understand. You have to do what you have to do.

FRED: ....

STEVE: Sure. No problem.

FRED: ....

STEVE: Good-bye.

WHAT ARE MOM AND DAD SAYING?

How about Steve's parents? What do you think they're saying to him? Fill in their "missing lines" and then practice the conversation with a friend.

STEVE: Hi, Mom and Dad! This is Steve.

MOM AND DAD: ....

STEVE: Yes, I'm fine.

MOM AND DAD: ....

STEVE: Listen, can I possibly come over for dinner on Sunday?

MOM AND DAD: ....

STEVE: WHAT do you have to do?

MOM AND DAD: ....

STEVE: Oh, I see.

WE CAN'T TALK NOW!

FINISH THE RAP!

| can't | have to | catch | talk | I | We | now | stop |

We _____¹ talk now.

We can't _____² now.

_____ _____³ talk now.

We can't talk _____⁴.

I can't talk now. I _____ _____⁵ go to work.

I _____ _____⁶ now. I have to catch a train.

I can't chat now. _____ _____ _____⁷ make a call.

I _____⁸ stop now. I have to _____⁹ a plane.

_____ _____¹⁰ talk now. _____ _____ _____¹¹ go to work.

We can't _____¹² now. We have to _____¹³ a train.

We can't chat _____¹⁴. _____¹⁵ have to make a call.

We _____¹⁶ stop now. We have to _____¹⁷ a plane.

We can't talk now.

WRITE YOUR OWN RAP!

We can't .... now.

I can't .... now. I have to ....

I can't .... now. I have to ....

I can't .... now. I have to ....

I can't .... now. I have to ....

We can't .... now. We have to ....

We can't .... now. We have to ....

We can't .... now. We have to ....

We can't .... now. We have to ....

We can't .... now.

SEGMENT 21

CAST PARTY

Linda can't come to the cast party. She wrote a note very quickly and made lots of mistakes. Can you find the mistakes and correct them?

Please Come to the Side By Side TV Cast Party

When: 7:00 P.M., Friday, December 15
Where: Side By Side TV Studio
Bring Family and Friends
R.S.V.P.

Dear Steve and Bill,

~~I~~ I'm very sorry, but I afraid John and I cant to come to you're party at Friday. John have to visits his parents in New York, and I have to working late.

Thanks because inviting us.

Sincerely,
Linda

WRITE YOUR OWN LETTER!

A friend invites YOU to a party. Write a letter to say you're sorry, but you can't go.

Please Come to a Birthday Party

For ....
When: ....
Where: ....
R.S.V.P.

Dear ....,
....
....
....
....
....

Sincerely,

....

FINISH THE SCRIPT!

Complete the conversation and then practice it with a friend.

A. Can you go to the movies on _____ (day of week) ?

B. I'm sorry. I can't. I have to _____ . How about _____ (day of week) ?

A. _____ (day of week) ? Let me see. No, I'm afraid I can't. I have to _____ . How about _____ (day of week) ?

B. _____ (day of week) ? Let me see.

THE NEXT LINE

Circle the best response.

1. How are you doing?
   - a.) Fine.
   - b. I'm cooking dinner.

2. Can you go shopping with me on Saturday?
   - a. I'm afraid.
   - b. I'm afraid not.

3. Sorry you can't come.
   - a. Me, too.
   - b. Yes, please.

4. See you later.
   - a. Bye.
   - b. Let me see.

5. Looks like I can't go to the movies with you this week.
   - a. How about tomorrow evening?
   - b. Maybe some other time.

6. I'm sorry I can't go to your party.
   - a. That's okay. I understand.
   - b. That's terrible.

WRONG LINE

Circle the expression that doesn't belong.

1. a. I'm sorry. I can't. b. I'm afraid not. c.) Of course I can. d. I'm afraid I can't.
2. a. Let me see. b. Definitely. c. I don't know. d. Let me check.
3. a. That's okay. b. I understand. c. No problem. d. That's terrible!
4. a. Fine. b. Sick. c. Great. d. Pretty good.
5. a. Bye. b. Hi. c. See you later. d. Have a nice day.
6. a. I'm sorry. b. That's too bad. c. Maybe some other time. d. That's wonderful!

SEGMENT 21

SEGMENT 21 SCRIPT

33:00 THEY CAN'T GO TO HERBERT'S PARTY

INTERVIEWER: Herbert is depressed. He's having a party today, but his friends can't go to his party. They're all busy.

INTERVIEWER: Can you go to Herbert's party?
FRIEND 1: No, I can't. I have to work.

INTERVIEWER: Can you go to Herbert's party?
FRIEND 2: No, we can't. We have to clean our house.

INTERVIEWER: Well, can your children go to Herbert's party?
FRIEND 3: No, they can't. They have to do their homework.

INTERVIEWER: Can you go to Herbert's party?
FRIEND 4: Yes, I can.
FRIEND 5: No, you can't. You have to go to the dentist.
(To Interviewer.)
She can't. She has to go to the dentist.
FRIEND 4: That's right. I can't.

INTERVIEWER: Can Michael go to Herbert's party?
FRIEND 6: No, he can't. He has to go to the doctor.

INTERVIEWER: Herbert?
HERBERT: Yes?
INTERVIEWER: I'm sorry, Herbert. Your friends can't come to your party. They all have things they have to do.

HERBERT: Oh. Well, can YOU come to my party?
INTERVIEWER: Me? Oh, no. I'm afraid I can't. I have to stay here and work. Sorry.
HERBERT: That's okay. I understand.

34:34 I'M SORRY. I CAN'T.

BOB: Oh hi, Steve! How are you today?
STEVE: Fine, Bob.
BOB: Hey, Steve, can you go to the ballgame with me tonight? I have two tickets.
STEVE: Tonight? Let me see. Gee, no, I can't. I have to go shopping.
BOB: Oh. That's too bad.
STEVE: But thanks for asking me.
BOB: Sure.

PATTY: Hi, Steve!
STEVE: Oh hi, Patty! How are you doing?
PATTY: Great. Listen, Steve, my roommate and I are having a party tomorrow evening. Can you come?
STEVE: Tomorrow evening? Let me check. Hmm. Looks like I can't. I have to do my laundry. But thanks for inviting me.
PATTY: Sorry you can't come.
STEVE: Me, too.
PATTY: See you later.
STEVE: See you later.

ALAN: Steve!
STEVE: Hi, Alan. How are you doing today?
ALAN: Pretty good. Listen, Steve, can you go sailing this Saturday?
STEVE: This Saturday? Hmm. I don't know. Nope. I'm afraid not. I have to clean my apartment this Saturday.
ALAN: Do you have to clean it on Saturday?
STEVE: Yes, I do. I really have to clean it this Saturday.
ALAN: Well, all right, Steve. See you later.
STEVE: See you later, Alan.

(Steve makes a telephone call.)

STEVE: Hi, Julie. This is Steve. Listen, Julie, can you have dinner with me on Sunday? You can't? You have to WHAT? Do you have to do that on Sunday? I see. Well, that's too bad. Maybe some other time. Bye.

(Steve makes another telephone call.)

STEVE: Hi, Fred. This is Steve. Great. Great. Listen, Fred, can you go bowling with me on Sunday? That's too bad. I'm sorry you can't. And you have to do that on Sunday? Of course I understand. You have to do what you have to do. Sure. No problem. Good-bye.

(Steve makes another telephone call.)

STEVE: Hi, Mom and Dad! This is Steve. Yes, I'm fine. Listen, can I possibly come over for dinner on Sunday? WHAT do you have to do? Oh, I see.

37:47

WE CAN'T TALK NOW!— GrammarRap

We can't talk now.
We can't talk now.
We can't talk now.
We can't talk now.

I can't talk now.
I have to go to work.
   I can't stop now.
   I have to catch a train.

I can't chat now.
I have to make a call.
   I can't stop now.
   I have to catch a plane.

We can't talk now.
We have to go to work.
We can't stop now.
We have to catch a train.

We can't chat now.
We have to make a call.
We can't stop now.
We have to catch a plane.

We can't talk now.
We can't talk now.
We can't talk now.
We can't talk now.

SEGMENT 21 SUMMARY

GRAMMAR

Can

Can	I / he / she / it / we / you / they	go to the party?

I / He / She / It / We / You / They	can / can't	go to the party.

Yes,	I / he / she / it / we / you / they	can.

No,	I / he / she / it / we / you / they	can't.

Have to

I / We / You / They	have to	work.
He / She / It	has to	

FUNCTIONS

Inquiring about Ability

Can *Michael go to Herbert's party?*

Expressing Ability

Yes, *I* can.

Expressing Inability

No, *I* can't.

Expressing Obligation

I have to *do my laundry.*
He has to *go to the doctor.*

Extending an Invitation

Can you *go to the ballgame with me tonight?*

Declining an Invitation

Sorry.
I'm sorry. I can't.
I'm afraid I can't.
Gee! No, I can't.
Looks like I can't.

I'm afraid not.

Expressing Regret

Sorry.
I'm sorry *you can't come.*

That's too bad.

Maybe some other time.

Greeting People

Hi! How are you today?
Hi! How are you doing?

Hesitating

Hmm.

Let me see.
Let me check.

I don't know.

Indicating Understanding

I see.

That's okay. I understand.
Of course I understand.

Sure.

No problem.

SEGMENT 22

- Plans and Intentions
- Future: Going to

♪ "We're gonna run, we're gonna walk, we're gonna sit, we're gonna talk . . . together Side by Side." ♪

PROGRAM LISTINGS

38:44 SBS-TV ON LOCATION
Two couples tell what they're going to do tomorrow.

39:38 PLANS FOR THE DAY
A college student tells where he's going today.

40:34 THEY'RE GOING TO THE BEACH
Mr. and Mrs. Brown are getting ready to go to the beach.

41:47 WHEN ARE YOU GOING TO WASH YOUR CLOTHES?
One roommate is upset because the other doesn't wash his clothes.

43:53 WHEN ARE YOU GOING TO CALL THE PLUMBER?
The sink is broken and it's time to call the plumber.

44:29 I'M GONNA BE VERY BUSY!
A "fast talker" tells his plans for the week.

46:05 HAPPY NEW YEAR!
Two people meet at a New Year's party and discover they have very similar plans for the coming year.

SBS-TV Backstage Bulletin Board

TO: Production Crew
Sets and props for this segment:

Dorm Room
 bed
 books
 bookbag
 ticket
 clothes
 headphones
 stereo

Bedroom
 skirt
 jacket
 pajamas
 suitcase
 sunglasses
 tie
 toothbrush

Party
 balloons
 decorations
 food
 glasses

Kitchen
 wrench
 sink

TO: Cast Members
Key words in this segment:

tomorrow morning/afternoon/evening/night
this week/month/year
next week/month/year

right now
immediately
at once
right away
tomorrow
today

SBS-TV ON LOCATION

SOUND CHECK

I'm	We're				
He's	You're	going to	fix	help	be
She's	They're		work	clean	
It's					

INTERVIEWER: Tell me, Alice, what are you going to do tomorrow?

ALICE: _____I'm going to work_____ ¹ in the yard.

INTERVIEWER: What's Fred going to do tomorrow?

ALICE: _____ ² his car.

FRED: Yes. _____ ³ my car,

and _____ ⁴ in the yard.

INTERVIEWER: What's the weather going to be like?

FRED: _____ ⁵ beautiful.

INTERVIEWER: Tell me, what are you going to do tomorrow?

HUSBAND: _____ ⁶ our house.

INTERVIEWER: I see.

HUSBAND: Yes. _____ ⁷ the living room,

and _____ ⁸ the basement.

WIFE: No, Harry. _____ ⁹ the basement,

and _____ ¹⁰ the living room.

_____ ¹¹ the basement.

INTERVIEWER: How about your children? Are they _____ ¹² help?

HUSBAND: Oh, yes. _____ ¹³ the attic.

INTERVIEWER: Well, happy cleaning!

39:38 PLANS FOR THE DAY

WHAT'S HAPPENING?

Complete these sentences.

_b_ 1. Today he's going to the a. tonight.
___ 2. He's going there b. library.
___ 3. He's going to his chemistry class c. concert ticket.
___ 4. He almost forgot his d. this morning.
___ 5. He's going to a concert e. this afternoon.
___ 6. Then they're going out to f. this evening.
___ 7. He looked in his wallet and he has his g. eat.
___ 8. He's going to be tired h. chemistry book.

EDITING MIX-UP

The video editor made a mistake! Put each pair of lines in the correct order.

1. ___ I'm going to the library.
   _1_ What are you going to do today?

2. ___ Are you going to the library this morning?
   ___ Yes. I'm going there right now.

3. ___ My chemistry book? Oh, yes! Thanks for reminding me.
   ___ Don't forget your chemistry book over there.

4. ___ Tell me, what are you going to do this evening?
   ___ I'm going to a concert with some friends.

5. ___ Yes. Here it is.
   ___ Do you have your concert ticket?

6. ___ That's for sure.
   ___ You know, you're going to be tired tonight.

SEGMENT 22

40:34 THEY'RE GOING TO THE BEACH

WHAT'S HAPPENING?

1. Mr. and Mrs. Brown are going to ____.
   a. Washington D.C.
   (b.) the beach

2. They're going there ____.
   a. tomorrow morning
   b. tomorrow afternoon

3. The reason is everybody else is going ____.
   a. tomorrow morning
   b. tomorrow afternoon

4. They're going to a fancy restaurant ____.
   a. tomorrow afternoon
   b. tomorrow evening

5. They're going to a hotel ____.
   a. tonight
   b. tomorrow night

THE NEXT LINE

Circle the correct response to each line in the scene.

1. Don't forget your sunglasses!
   (a.) They're right here.
   b. There they are.

2. I'm going to wear this at the restaurant tomorrow evening. What do you think?
   a. It's fine.
   b. It's fun.

3. Do you like this jacket and tie?
   a. It's purple.
   b. It's perfect.

4. Don't forget your pajamas and toothbrush!
   a. I have them right here.
   b. Are they here?

5. Well, I think that's everything.
   a. I don't know.
   b. I think so.

6. We're going to have a great time at the beach tomorrow!
   a. The weather's going to be beautiful!
   b. It's going to be wonderful!

WHOSE LINE?

a. "I'm going to wear this at the restaurant tomorrow evening." (Helen) Howard

b. "Do you like this jacket and tie?" Helen Howard

c. "Don't forget your pajamas and toothbrush!" Helen Howard

d. "I have them right here." Helen Howard

e. "Well, I think that's everything." Helen Howard

PICTURE THIS!

Match the lines above with the following scenes from the video.

1 a   2 ____   3 ____

4 ____   5 ____

CLOSE-UP

You're on Side by Side TV! Tell the viewers: What are YOU going to do tomorrow morning? tomorrow afternoon? tomorrow evening? tomorrow night?

....

....

....

....

SEGMENT 22

WHEN ARE YOU GOING TO WASH YOUR CLOTHES?

WHOSE LINE?

1	"When are you going to wash your clothes?"	Lance	(Theodore)
2	"Are you really going to wash them this week?"	Lance	Theodore
3	"Well, maybe next week."	Lance	Theodore
4	"Next week?! How about this week?"	Lance	Theodore
5	"There's no reason to get angry."	Lance	Theodore
6	"I promise."	Lance	Theodore
7	"This month?!"	Lance	Theodore
8	"I don't think you're going to wash them next month."	Lance	Theodore
9	"I'm going to wash your clothes!"	Lance	Theodore
10	"You're going to wash MY clothes?!"	Lance	Theodore
11	"You're NEVER going to wash them!"	Lance	Theodore
12	"Next year, I'm going to wash YOUR clothes!"	Lance	Theodore

EDITING MIX-UP

The video editor made a mistake! Put the following lines in the correct order.

____ Lance, are you really going to wash them this week?

____ Now come on, Theodore, calm down! There's no reason to get angry. I'm going to wash my clothes sometime this month. I promise.

__1__ Lance, when are you going to wash your clothes?

____ Well, maybe next week.

____ You're right! This IS a very busy month. I'm going to wash my clothes NEXT month. And that's a promise, Theodore.

____ I'm going to wash them this week.

____ Next week?! How about this week?

____ Next month?! Lance, I don't believe you!

____ This month?! Lance, I don't believe this! Are you serious?!

SEGMENT 22

43:53 WHEN ARE YOU GOING TO CALL THE PLUMBER?

EDITING MIX-UP

The video editor made a mistake! Put the following lines in the correct order.

____ I'm fixing the sink.

____ Sure I can!

____ Yes!

____ YOU'RE fixing the sink?!

__1__ Edward, what are you doing?

____ Edward, come on! You can't fix the sink!

SOUND CHECK

| right away | right now | immediately | at once |

When are you going to call the plumber?

I'm going to call him __right now__ 1.

Edward, I think we need a plumber _____ 2.

_____ 3!!

You're right. I'm going to call the plumber _____ 4!

CONTINUE THE SCENE!

Edward is calling the plumber. Complete their telephone conversation and practice it with a friend.

EDWARD: ....

PLUMBER: ....

EDWARD: ....

PLUMBER: ....

SEGMENT 22

44:29 I'M GONNA BE VERY BUSY!

LISTENING CHALLENGE!

This person speaks very fast! Watch the scene several times and decide whether the following statements about his busy week are true or false.

Next Sunday:

1. He's going to visit his mother and father. (True) False
2. His mother is going to make soup. True False
3. His father is going to cook in the kitchen. True False
4. His parents are going to ask him about his job. True False
5. He's going to ask them about their friends. True False
6. They're going to have a good time. True False

SUNDAY

Next Monday:

7. He's going to go to work. True False
8. He's going to drive his car to work. True False
9. The mechanic is going to fix his car. True False
10. He's going to walk to work. True False
11. He's going to get his car after work. True False
12. It's going to be cheap to fix his car. True False
13. He's going to be upset. True False

MONDAY

Next Tuesday:

14. He's going to get to work late. True False
15. He's going to ask his boss for a raise. True False
16. His boss is going to give him a raise. True False
17. He isn't going to be happy. True False

TUESDAY

Next Wednesday:

18. He's going to fly to Chicago. True False
19. He's going to go to a museum in the morning. True False
20. He's going to have lunch alone. True False
21. He's going to a meeting in the afternoon. True False
22. He's going to drive home in the evening. True False
23. He's going to be very tired. True False

WEDNESDAY

SEGMENT 22

Next Thursday:

24 He's going to write a report. True False

25 He's going to write about the lunch. True False

26 He's going to type the report and give it to his boss. True False

Next Friday:

27 He's going to go to the dentist before work. True False

28 The dentist is going to clean his teeth. True False

29 He's going to be nervous, but the dentist isn't going to hurt him. True False

30 He likes his dentist and he believes what his dentist says. True False

Next Saturday:

31 He's going to relax and have fun. True False

32 He's going to go bowling after he goes jogging. True False

33 He's going to have lunch at home. True False

34 He's going to go sailing in the afternoon. True False

35 Then he's going to go shopping. True False

36 He's going to see a play and go dancing in the evening. True False

CLOSE-UP

You're on Side by Side TV! Write about YOUR busy week next week. Then tell a friend about your busy week as fast as you can!

I'm going to be very busy next week. It's going to be a very busy week. Do you want to know how busy I'm going to be? I'm going to tell you right now.

....

....

....

....

....

Yes, it's going to be a very busy week, but I guess I like it that way. Oh, I'm going to be late. I'm going to go now. Bye!

HAPPY NEW YEAR!

SCENE CHECK

1. He's / **She's** going to start a new job in January.
2. He's / She's going to start a new job in February.
3. In March / April he's going to California to visit his brother / mother.
4. In March / April she's going there to visit her niece / sister.
5. She's / He's going to buy a new car in May.
6. She's / He's going to buy a new car in June.
7. He's / She's going to Canada in June / July.
8. She's going there in July / August. She usually / always goes there then.
9. He's / She's going to acting school in September.
10. He's / She's going to acting school in October.
11. She's getting married in November / December.
12. He is / isn't going to get married in December.

SCENE REVIEW

Which of the months in this scene have these associations?

1. Months associated with school: _September, October_
2. Months associated with work: _____
3. Months associated with travel: _____

What do you think?

4. His favorite month is going to be _____ because _____
5. Her favorite month is going to be _____ because _____

PICTURE THIS!

Match these lines with the appropriate scenes below.

a. "Well, just two more minutes and it's a brand new year."
b. "Next year's going to be a great year!"
c. "I'm going to California in April to visit my sister."
d. "I'm going to buy a new car in May."
e. "You're going to WHAT?"
f. "Happy New Year!"

1. _b_   2. ___   3. ___

4. ___   5. ___   6. ___

CLOSE-UP

You're on Side by Side TV! Tell about your plans for next year.

I'm really looking forward to next year!

In January _____   In July _____

In February _____   In August _____

In March _____   In September _____

In April _____   In October _____

In May _____   In November _____

In June _____   In December _____

Next year is really going to be an exciting year!

SEGMENT 22

WHAT'S THE QUESTION?

Bob is going away for the weekend. His friend Peter is asking him about his trip.

Who	What	When	Where	Why

1. When are you going to _____ leave? — Tomorrow morning.
2. _____ stay? — At a small hotel.
3. _____ stay there? — Because I like small hotels.
4. _____ go with? — My cousins Tim and Philip.
5. _____ do there? — We're going to see the sights.

WHAT'S MY LINE?

1. Juan is (going / go) to visit his parents this weekend.

2. Noriko is going to (have / has) dinner with her friends this (tonight / evening).

3. Gary, (is / are) your roommate going to clean your apartment this (weekend / night)?

4. When (is / are) you going to start your new job at the factory?

5. Gregory! Finish your homework right (immediately / now)!

6. Our front door is broken. When (are you going to / you are going to) call the superintendent?

7. My brother and I are (going to / going to go) Mexico this year.

8. What (are / am) I going to cook for dinner tonight? We don't have any food in the refrigerator!

9. I'm going to call the police at (away / once)!

10. You're (going get / going to get) married next month? Congratulations!

SEGMENT 22 SCRIPT

38:44 SBS-TV ON LOCATION

INTERVIEWER: Hi, folks!
HUSBAND 1: Oh hello.
WIFE 1: Hi.
INTERVIEWER: What are your names?
WIFE 1: I'm Alice, and he's Fred.
INTERVIEWER: Tell me, Alice, what are you going to do tomorrow?
WIFE 1: I'm going to work in the yard.
INTERVIEWER: What's Fred going to do tomorrow?
WIFE 1: He's going to fix his car.
HUSBAND 1: Yes. I'm going to fix my car, and she's going to work in the yard.
INTERVIEWER: What's the weather going to be like?
HUSBAND 1: It's going to be beautiful.
INTERVIEWER: Well, have a nice day!
WIFE 1: Thanks.
HUSBAND 1: You, too.

INTERVIEWER: Hello.
HUSBAND 2: Hi.
WIFE 2: Hello.
INTERVIEWER: Tell me, what are you going to do tomorrow?
HUSBAND 2: We're going to clean our house.
INTERVIEWER: I see.
HUSBAND 2: Yes. I'm going to clean the living room, and she's going to clean the basement.
WIFE 2: No, Harry. YOU'RE going to clean the basement, and I'M going to clean the living room. *(To Interviewer.)* HE'S going to clean the basement.
INTERVIEWER: How about your children? Are they going to help?
HUSBAND 2: Oh, yes. They're going to clean the attic.
INTERVIEWER: Well, happy cleaning!
HUSBAND 2: Thanks.
WIFE 2: Bye.
HUSBAND 2: Who IS that?
WIFE 2: I don't know.

39:38 PLANS FOR THE DAY

INTERVIEWER: What are you going to do today?
STUDENT: I'm going to the library.
INTERVIEWER: Are you going to the library this morning?
STUDENT: Yes. I'm going there right now.
INTERVIEWER: What are you going to do this afternoon?
STUDENT: I'm going to my chemistry class.
INTERVIEWER: I see. Don't forget your chemistry book over there.
STUDENT: My chemistry book? Oh, yes! Thanks for reminding me.
INTERVIEWER: You're welcome. Tell me, what are you going to do this evening?
STUDENT: I'm going to a concert with some friends, and then we're all going out for a late dinner.
INTERVIEWER: Do you have your concert ticket?
STUDENT: Yes. Here it is.
INTERVIEWER: You know, you're going to be tired tonight.
STUDENT: That's for sure.

INTERVIEWER: Well, have a nice day, and enjoy the concert!
STUDENT: Thanks.

⏱ 40:34 THEY'RE GOING TO THE BEACH

PERSON 1: What are Mr. and Mrs. Brown going to do tomorrow?
PERSON 2: They're going to the beach.
MRS. BROWN: Don't forget your sunglasses!
MR. BROWN: They're right here.
PERSON 1: Are they going to the beach tomorrow morning?
MR. BROWN: You know, Helen, everybody's going to the beach tomorrow morning.
MRS. BROWN: You're right, Howard. Let's go tomorrow afternoon.
PERSON 2: No, I don't think so. I think they're going to the beach tomorrow afternoon.
PERSON 1: What are they going to do tomorrow evening?
MRS. BROWN: I'm going to wear this at the restaurant tomorrow evening. What do you think?
MR. BROWN: It's fine. Do you like this jacket and tie?
MRS. BROWN: It's perfect.
PERSON 2: They're going to a fancy restaurant near the beach.
PERSON 1: Are they going back home tomorrow night?
MRS. BROWN: Don't forget your pajamas and toothbrush!
MR. BROWN: I have them right here.
PERSON 2: No. They're going to a hotel.
MR. BROWN: Well, I think that's everything.

MRS. BROWN: I think so.
MR. BROWN: We're going to have a great time at the beach tomorrow!
MRS. BROWN: It's going to be wonderful!

⏱ 41:47 WHEN ARE YOU GOING TO WASH YOUR CLOTHES?

THEODORE: Lance! Lance!
LANCE: Hey! What's going on?
THEODORE: Lance, when are you going to wash your clothes?
LANCE: I'm going to wash them this week.
THEODORE: Lance, are you really going to wash them this week?
LANCE: Well, maybe next week.
THEODORE: Next week?! How about this week?
LANCE: Now come on, Theodore, calm down! There's no reason to get angry. I'm going to wash my clothes sometime this month. I promise.
THEODORE: This month?! Lance, I don't believe this! Are you serious?!
LANCE: You're right! This IS a very busy month. I'm going to wash my clothes NEXT month. And that's a promise, Theodore.
THEODORE: Next month?! Lance, I don't believe you! I don't think you're going to wash them next month. In fact, I don't think you're going to wash them this year!
LANCE: Whoa! What are you doing?

SEGMENT 22

THEODORE: What do you think I'm doing? I'm going to wash your clothes!
LANCE: You're going to wash MY clothes?!
THEODORE: Yes, I'm going to wash your clothes, because you're not going to wash them! I know it! You're NEVER going to wash them!
LANCE: Gee, Theodore, I don't know what to say.
THEODORE: Don't say anything.
LANCE: Tell you what, Theodore. I'm going to make you a promise. Next year I'm going to wash YOUR clothes!
THEODORE: You're going to what?!
LANCE: Next year, I'm going to wash YOUR clothes! And that's a promise.
THEODORE: Sure, Lance!
LANCE: It's a promise!

43:53 WHEN ARE YOU GOING TO CALL THE PLUMBER?

WIFE: Edward, what are you doing?
HUSBAND: I'm fixing the sink.
WIFE: YOU'RE fixing the sink?!
HUSBAND: Yes!
WIFE: Edward, come on! You can't fix the sink!
HUSBAND: Sure I can!
WIFE: When are you going to call the plumber?
HUSBAND: I'm going to call him right now.

WIFE: Edward, I think we need a plumber immediately. At once!!
HUSBAND: You're right. I'm going to call the plumber right away!

44:29 I'M GONNA BE VERY BUSY!

FAST TALKER: I'm gonna be very busy next week. It's gonna be a very busy week. Do you wanna know how busy I'm gonna be? I'm gonna tell you right now.

Next Sunday I'm gonna visit my mother and father. My mother's gonna make soup and my father's gonna cook chicken on the barbecue. They're gonna ask me about my work. I'm gonna ask them about their friends. We're gonna have a very nice time.

Next Monday I'm gonna go to work. I'm not gonna drive my car to work because I'm gonna bring my car to the mechanic and he's gonna fix it. I'm gonna take the bus to work. After work I'm gonna take the bus to the garage and get my car. The mechanic's gonna give me my bill and I'm gonna be upset because it's gonna cost a lot of money.

Next Tuesday I'm gonna get to work early because I'm gonna talk to my boss. I'm gonna ask my boss for a raise. She's gonna say "No, maybe next year" and I'm gonna say "Okay" and walk out and I'm gonna be sad.

Next Wednesday I'm gonna fly to Chicago. I'm gonna go to a meeting in the morning, I'm gonna have lunch with some people, and I'm gonna go to another meeting in the afternoon. I'm gonna fly home in the evening, and I'm gonna be very tired.

Next Thursday I'm gonna write a report about my meetings in Chicago. I'm not gonna write about the lunch because the lunch isn't gonna be very good. I'm gonna type the report on my computer, I'm gonna print it on my printer, and I'm gonna give it to my boss.

Next Friday I'm not gonna go to work right away. I'm gonna go to the dentist. My dentist is gonna look at my teeth and his assistant's gonna clean them. I'm gonna be a little nervous, but my dentist is gonna tell me, "Don't worry. I'm not gonna hurt you. Everything's gonna be okay." And I'm gonna believe him because my dentist is a very nice person.

Next Saturday I'm gonna relax and have fun. I'm gonna go jogging. Then I'm gonna go bowling. I'm gonna have lunch at my favorite restaurant. Then in the afternoon, I'm gonna go sailing. Then I'm gonna go shopping. In the evening, I'm gonna go to a movie with some friends and then we're gonna go dancing.

Yes, it's gonna be a very busy week, but I guess I like it that way. Oh, I'm gonna be late. I'm gonna go now. Bye!

46:05

HAPPY NEW YEAR!

MAN: Well, just two more minutes and it's a brand new year.
WOMAN: Just two more minutes. How about that!
MAN: This sure is a great New Year's party.
WOMAN: It sure is. It's a very nice party.
BOTH: Are you looking forward to . . .
MAN: I'm sorry. Go ahead.
WOMAN: Are you looking forward to next year?
MAN: Yes, I am. How about you?
WOMAN: Me, too. Next year's going to be a great year! In January I'm going to start a new job.
MAN: That's interesting. I'm going to start a new job in February.
WOMAN: How about that!
MAN: In March I'm going to California to visit my brother.
WOMAN: Really? I'm going to California in April to visit my sister.
MAN: Well, how do you like that!
WOMAN: May is going to be an exciting month. I'm going to buy a new car in May.

MAN: What a coincidence! I'm going to buy a new car in June!
WOMAN: I can't believe it!
MAN: It's amazing!
WOMAN: Are you going to go on a summer vacation next year?
MAN: Yes. I'm going to Canada in July. Don't tell me you're going to Canada, too!
WOMAN: I sure am. I'm going to go there in August. I always go to Canada in August . . . every year.
MAN: September is going to be a very important month. In September I'm going to acting school.
WOMAN: No! You're kidding! I'm going to acting school in October!

MAN: That's incredible! So what's going to happen with you in November?
WOMAN: November? Well . . . uh . . . I'm going to get married in November.
MAN: You're going to WHAT?
WOMAN: I'm going to get married in November.
MAN: Oh.
WOMAN: I suppose you're going to get married in December, right?
MAN: December? Uh . . . no . . . I'm not going to get married in December.
WOMAN: Look at the time! It's twelve o'clock! Happy New Year!
EVERYONE: Happy New Year!
MAN: Happy New Year!

SEGMENT 22 SUMMARY

GRAMMAR

Future: Going to

What	am	I	going to do?
	is	he / she / it	
	are	we / you / they	

(I am)	I'm	going to read.
(He is)	He's	
(She is)	She's	
(It is)	It's	
(We are)	We're	
(You are)	You're	
(They are)	They're	

Time Expressions

I'm going to wash my clothes	today. this morning. this afternoon. this evening. tonight.	tomorrow. tomorrow morning. tomorrow afternoon. tomorrow evening. tomorrow night.	right now. right away. immediately. at once.

I'm going to fix my car	this/next	week. month. year. spring. summer. fall (autumn). winter.	Sunday. Monday. Tuesday. Wednesday. Thursday. Friday. Saturday.	January. February. March. April. May. June.	July. August. September. October. November. December.

FUNCTIONS

Inquiring about Intention

What are you going to do *tomorrow*?
When are you going to *wash your clothes*?

Expressing Intention

I'm going to *paint my kitchen*.
I'm going to *wash my clothes this week*.

Asking for Information

Tell me, _____?

Expressing Good Wishes

Have a nice *day*!
Enjoy the *concert*!

SEGMENT 23

- **Weather Forecasts**
- **Telling Time**
- **Future: Going to**
- **Want to**

"What's the forecast? What's the weather? Rain or shine, we'll be together... Side by Side."

PROGRAM LISTINGS

48:49 WORLD WEATHER FORECAST
Meteorologist Maria Hernandez gives the weather forecast for places around the world.

50:42 WHAT'S THE FORECAST?
Two college students talk after class.

51:06 SBS-TV ON LOCATION
People tell the time.

51:31 WHAT TIME IS IT?
A wife is upset that she and her husband are going to be late for a movie.

51:56 SBS-TV ON LOCATION
People give information about schedules.

52:43 IT'S TIME FOR BED
An elderly man falls asleep in front of the TV.

53:22 TO BE WITH YOU—Music Video
A singer promises to be faithful through all the seasons, months of the year, and other time expressions.

SBS-TV Backstage Bulletin Board

TO: Production Crew
Sets and props for this segment:

TV Studio
weather
map

Classroom
desks
map
books

Bathroom
mirror
sink
razor
shaving cream

Sports Stadium
tickets

Train Station
tickets
suitcase

Movie Theater
tickets

Living Room
TV
sofa

Stage
gold records

TO: Cast Members
Key words in this segment:

forecast
clear
drizzle
foggy
a quarter to
a quarter after
half past
midnight
noon
What time . . . ?

begin
leave
go swimming
hope
hurry
shave
wait
want
late

WORLD WEATHER FORECAST

YES OR NO?

		Yes	No
1	It's going to be sunny in Puerto Rico.	(Yes)	No
2	It's going to be clear in Venezuela.	Yes	No
3	It's going to rain in France.	Yes	No
4	It's just going to rain a little in Poland.	Yes	No
5	It's going to snow in Ukraine.	Yes	No
6	It's going to be cloudy in Morocco.	Yes	No
7	It isn't going to be clear in Portugal.	Yes	No
8	It isn't going to be cold in Saudi Arabia.	Yes	No
9	It's going to be warm in Thailand.	Yes	No
10	It's going to be cool in Korea.	Yes	No
11	Wear a light jacket in Lithuania.	Yes	No

A GOOD DAY FOR THE BEACH

Put a check (✔) next to the countries where the weather is going to be good for a day at the beach.

✔ Puerto Rico	____ Poland	____ Portugal	____ Korea
____ Venezuela	____ Ukraine	____ Saudi Arabia	____ Lithuania
____ France	____ Morocco	____ Thailand	

WEATHER CHALLENGE!

Watch the weather forecast for these 19 countries several times. How many answers can you give?

1	Afghanistan:	(cool)	cold	11	Indonesia:	cool	hot
2	Australia:	rain	warm	12	Italy:	rain	snow
3	Bolivia:	hot	foggy	13	Japan:	clear	drizzle
4	Canada:	sunny	snow	14	Malaysia:	warm	hot
5	Chile:	cool	clear	15	Mexico:	raining	sunny
6	Colombia:	cloudy	sunny	16	Panama:	hot	foggy
7	Ecuador:	hot	drizzle	17	Peru:	rain	clear
8	El Salvador:	warm	cool	18	Taiwan:	cold	cool
9	Estonia:	cloudy	foggy	19	United States:	snow	showers
10	Guatemala:	warm	rain				

ON CAMERA

You're on Side by Side TV! Give tomorrow's weather forecast for places around the world.

Hello, everybody. Let's take a look at tomorrow's weather forecast around the world.

It's going to ....

....

....

....

....

....

And that's the World Weather Forecast.

I'm .... Have a
      (your name)
nice day.

WHAT'S THE FORECAST?

EDITING MIX-UP

The video editor made a mistake! Put the following lines in the correct order.

____ The radio says it's going to rain.

____ What's the forecast?

__1__ What are you going to do tomorrow?

____ I hope you're right. I REALLY want to go swimming.

____ That's strange! According to the newspaper, it's going to be sunny.

____ I don't know. I want to go swimming, but I think the weather is going to be bad.

SBS-TV ON LOCATION

SOUND CHECK

| it | time | what | can | have | eleven | forty-five |
| it's | o'clock | what's | do | tell | fifteen | thirty |

A. Excuse me. __What__¹ time is _____²?

B. It's _____³ o'clock.

A. Pardon me. Do you _____⁴ the time?

B. Yes. _____⁵ eleven _____⁶.

A. _____⁷ the _____⁸?

B. It's eleven _____⁹.

SEGMENT 23

A. Excuse me. _____10_____ you _____11_____ me the time?

B. Certainly. It's eleven _____12_____.

A. What time _____13_____ you _____14_____?

B. It's twelve _____15_____.

WHAT'S THE TIME?

The interviewer is asking some other people the time. Draw the correct time on the clocks.

1. A. Excuse me. What time is it?
   B. It's two o'clock.

2. A. Excuse me. What's the time?
   B. It's three thirty.

3. A. Can you please tell me the time?
   B. Sure. It's four forty-five.

4. A. Pardon me. What time do you have?
   B. It's five fifteen.

5. A. Excuse me. Do you have the time?
   B. Yes. It's six o'clock.

6. A. What time do you have?
   B. It's seven forty-five.

SEGMENT 23

51:31 WHAT TIME IS IT?

EDITING MIX-UP

The video editor made a mistake! Put the following lines in the correct order.

___ Why? What time is it?

___ It begins at eight o'clock.

___ It's seven thirty! We have to leave right now!

___ Please try to hurry! I don't want to be late for the movie.

_1_ What time does the movie begin?

___ Oh, but I can't leave now. I'm shaving!

___ Oooh!

___ At eight o'clock? Oh, no! We're going to be late!

WHOSE LINE?

1	"Oh, no! We're going to be late!"	Husband	(Wife)
2	"Oh, but I can't leave now. I'm shaving!"	Husband	Wife
3	"It's seven thirty! We have to leave right now!"	Husband	Wife
4	"Please try to hurry! I don't want to be late for the movie."	Husband	Wife
5	"Why? What time is it?"	Husband	Wife
6	"Oooh!"	Husband	Wife

SEGMENT 23

51:56 SBS-TV ON LOCATION

YES, NO, OR MAYBE?

1	The game begins at a quarter after two.	(Yes)	No	Maybe
2	The game begins at two fifty.	Yes	No	Maybe
3	It's a baseball game.	Yes	No	Maybe
4	The game begins at two fifteen.	Yes	No	Maybe
5	The train to New York is Amtrak Train #64.	Yes	No	Maybe
6	The train to New York is in the station.	Yes	No	Maybe
7	The train to New York leaves at five thirty.	Yes	No	Maybe
8	The train to New York leaves at half past five.	Yes	No	Maybe
9	*The Flowers in Priscilla's Garden* is a movie.	Yes	No	Maybe
10	*The Flowers in Priscilla's Garden* begins at a quarter to eight.	Yes	No	Maybe
11	*The Flowers in Priscilla's Garden* starts at seven forty-five.	Yes	No	Maybe
12	*Space Wars* begins at a quarter after eight.	Yes	No	Maybe
13	Everybody is buying tickets for *The Flowers in Priscilla's Garden*.	Yes	No	Maybe

SEGMENT 23

CONTINUE THE SCENE AT THE TRAIN STATION!

More people at the train station need information. Using the schedule below, ask and answer questions about the different trains.

Train Number	Destination	Leaves
54	New York	5:30
63	Chicago	6:15
72	Philadelphia	7:30
84	Miami	8:45
90	Cleveland	9:15
95	Los Angeles	10:45

A. What time does the train to _____ leave?

B. It leaves at (a quarter after/a quarter to/half past) _____.

A. (A quarter after/A quarter to/Half past) _____? Thanks.

CONTINUE THE SCENE AT THE MOVIE THEATER!

More people at the movie theater need information. Using the schedule below, ask and answer questions about the different movies.

Movies

THE FLOWERS IN PRISCILLA'S GARDEN	7:45
SPACE WARS	8:15
A HUNDRED BALLOONS	8:30
MURDER AT MIDNIGHT	9:45
THE LAST GOOD-BYE	10:30

A. What time does _____ _____ start?

B. It begins at _____.

A. _____? Thanks.
    (time)

52:43 IT'S TIME FOR BED

EDITING MIX-UP

The video editor made a mistake! Put the following lines in the correct order.

____ Twelve noon?

____ Yes, dear. It's time for bed.

____ Twelve midnight?!

_1_ What time is it?

____ No, don't be silly! It's nighttime. It's twelve midnight!

____ It's twelve o'clock, dear.

____ Oh.

53:22 TO BE WITH YOU

FINISH THE SONG!

I'm	be	after	day	fall	right	April
it's	wait	in	month	summer		December
you	waiting	past	week			February
		to	year			July
		with				September

Any day, any __week__ ¹, any month, any _____ ²,

I'm gonna _____ ³ right here to be with you.

_____ ⁴ the spring, in the _____ ⁵,

in the winter, or the _____ ⁶. Just call.

I'm _____ ⁷ here to be with you.

_____ ⁸ gonna wait from January, _____ ⁹, March,

_____ ¹⁰, May, June and _____ ¹¹, August, _____ ¹², October

and November, and all of _____ ¹³. I'm going to wait . . .

_____ ¹⁴ one o' clock, a quarter _____ ¹⁵. It's half _____ ¹⁶ one, a quarter

_____ ¹⁷ two. And I'm gonna _____ ¹⁸ right here to be with you.

Any _____ ¹⁹, any week, any _____ ²⁰, any year,

I'm gonna wait _____ ²¹ here to be _____ ²² you.

Yes, I'm gonna wait right here _____ ²³ with _____ ²⁴.

WHAT'S NEXT?

The rock singer is composing a new song. Can you help?

1. "I think of you in January, February, and __March__."

2. "I'm yours in May, June, and _____."

3. "I want to be with you in the morning, afternoon, and _____."

4. "I need you every Monday, Tuesday, and _____."

5. "I love you in the summer, fall, and _____."

WRAP-UP

WHAT ARE THEY SAYING?

do	begin	it	a quarter after	eight	midnight	at
does	begins	it's	a quarter to	three		time
is	leaves		half past	two		what
				twelve		

1. A. __What__ time __does__ the train to Chicago leave?
   B. It __leaves__ at __half__ __past__ __three__.

3. A. _____ _____ the movie begin?
   B. It _____ at _____ _____ _____.

2. A. Excuse me. What time _____ _____?
   B. It's _____ _____ _____.

4. A. What's the _____?
   B. It's _____ o'clock, _____. Time for bed, dear.

5. A. _____ you have the time?
   B. Yes. _____ _____ _____ _____ _____.

6. A. _____ _____ does the ballgame _____?
   B. _____ begins at _____ _____ _____.

SEGMENT 23 SCRIPT

48:49 WORLD WEATHER FORECAST

ANNOUNCER: Now here's the World Weather Forecast from Side by Side TV News, with Side by Side meteorologist, Maria Hernandez.

MARIA HERNANDEZ: Hello, everybody. Let's take a look at tomorrow's weather forecast around the world.

It's going to be sunny in Puerto Rico. It's going to be cloudy in Venezuela. All our friends in France, get your umbrellas ready! It's going to rain. It isn't going to rain very hard in Poland. It's just going to drizzle. It's going to snow in Ukraine. It's going to be clear in Morocco. And be careful driving in Portugal. It's going to be foggy.

And looking at some temperatures around the world: It's going to be hot in Saudi Arabia. It's going to be warm in Thailand. Wear a light jacket in Korea. It's going to be cool. And put on a heavy coat in Lithuania. It's going to be cold! Now here's the weather forecast for other parts of the world.

(Weather around the world.)

Afghanistan:	cool
Australia:	warm
Bolivia:	hot
Canada:	snow
Chile:	clear
Colombia:	sunny
Ecuador:	hot
El Salvador:	warm
Estonia:	foggy
Guatemala:	warm
Indonesia:	hot
Italy:	rain
Japan:	drizzle
Malaysia:	warm
Mexico:	sunny
Panama:	hot
Peru:	clear
Taiwan:	cool
United States:	snow

MARIA HERNANDEZ: And that's the World Weather Forecast from Side by Side TV News. I'm Maria Hernandez. Have a nice day.

50:42 WHAT'S THE FORECAST?

STUDENT 1: What are you going to do tomorrow?
STUDENT 2: I don't know. I want to go swimming, but I think the weather is going to be bad.
STUDENT 1: What's the forecast?
STUDENT 2: The radio says it's going to rain.
STUDENT 1: That's strange! According to the newspaper, it's going to be sunny.

STUDENT 2: I hope you're right. I REALLY want to go swimming.

51:06 SBS-TV ON LOCATION

INTERVIEWER: Excuse me. What time is it?
PERSON 1: It's eleven o'clock.

INTERVIEWER: Pardon me. Do you have the time?
PERSON 2: Yes. It's eleven fifteen.

INTERVIEWER: What's the time?
PERSON 3: It's eleven thirty.

INTERVIEWER: Excuse me. Can you tell me the time?
PERSON 4: Certainly. It's eleven forty-five.

INTERVIEWER: What time do you have?
PERSON 5: It's twelve o'clock.

51:31 WHAT TIME IS IT?

WIFE: What time does the movie begin?
HUSBAND: It begins at eight o'clock.
WIFE: At eight o'clock? Oh, no! We're going to be late!
HUSBAND: Why? What time is it?
WIFE: It's seven thirty! We have to leave right now!
HUSBAND: Oh, but I can't leave now. I'm shaving!
WIFE: Please try to hurry! I don't want to be late for the movie.

(The husband tries to hurry and cuts himself.)
HUSBAND: Oooh!

51:56 SBS-TV ON LOCATION

INTERVIEWER: Excuse me. What time does the game begin?
PERSON 1: It begins at a quarter after two.
INTERVIEWER: A quarter after two?
PERSON 1: Yes. That's right.
INTERVIEWER: Thanks.

ANNOUNCEMENT: This is the final boarding call for Amtrak Train number fifty-four to New York. Train number fifty-four. All aboard, please.
INTERVIEWER: Pardon me. Are you on the train to New York?
PERSON 2: Yes, I am.
INTERVIEWER: What time does the train leave?
PERSON 2: It leaves at half past five.
INTERVIEWER: Half past five? Thanks.

INTERVIEWER: Excuse me. What time does *The Flowers in Priscilla's Garden* start?

PERSON 3: *The Flowers in Priscilla's Garden?* I have no idea. I'm seeing *Space Wars.* Let me take a look. It begins at a quarter to eight.
INTERVIEWER: A quarter to eight? Thanks.

53:22 TO BE WITH YOU—Music Video

Any day, any week,
Any month, any year,
I'm gonna wait right here
To be with you.

In the spring, in the summer,
In the winter, or the fall,
Just call. I'm waiting here
To be with you.

I'm gonna wait from
January, February,
March, April, May,
June and July,
August, September,
October and November,
And all of December.
I'm gonna wait . . .

It's one o'clock, a quarter after,
It's half past one, a quarter to two,
And I'm gonna wait right here
To be with you.

Any day, any week,
Any month, any year,
I'm gonna wait right here
To be with you.

Yes, I'm gonna wait right here
To be with you.

52:43 IT'S TIME FOR BED

HUSBAND: What time is it?
WIFE: It's twelve o'clock, dear.
HUSBAND: Twelve noon?
WIFE: No, don't be silly! It's nighttime. It's twelve midnight!
HUSBAND: Twelve midnight?!
WIFE: Yes, dear. It's time for bed.
HUSBAND: Oh.

SEGMENT 23 SUMMARY

GRAMMAR

Future: Going to

What	am	I	going to do?
	is	he / she / it	
	are	we / you / they	

(I am)	I'm	going to read.
(He is)	He's	
(She is)	She's	
(It is)	It's	
(We are)	We're	
(You are)	You're	
(They are)	They're	

Time Expressions

It's	11:00 (eleven o'clock).
	11:15 (eleven fifteen/a quarter after eleven).
	11:30 (eleven thirty/half past eleven).
	11:45 (eleven forty-five/a quarter to twelve).
	noon.
	midnight.

Want to

I / We / You / They	want to	study.
He / She / It	wants to	

FUNCTIONS

Inquiring about Intention

What are you going to do *tomorrow*?

Expressing Want–Desire

I want to *go swimming*.
I really want to *go swimming*.

I hope *you're right*.

I don't want to *be late*.

Leave Taking

Have a nice day.

Asking for and Reporting Information

What's the forecast?
  The radio says it's going to *rain*.
  According to the newspaper, it's going to *be sunny*.

What time is it?
What's the time?
What time do you have?
Do you have the time?

What time does the *train* leave?
  It leaves at *5:30*.

SEGMENT 24

- "Aches and Pains"
- Past Tense: Regular and Irregular Verbs

🎵 "We cooked and played and then we talked. We ate and drank and then we walked . . . Side by Side."

PROGRAM LISTINGS

55:15 HOW DO YOU FEEL TODAY?
Bob, an office worker, feels worse and worse as he eats throughout the workday.

56:45 SBS-TV ON LOCATION
People tell about their ailments.

57:07 SBS-TV ON LOCATION
People tell what they did yesterday.

57:28 WHAT'S THE MATTER?
A mother is concerned about her daughter's husband.

57:47 SBS-TV ON LOCATION
People tell why they aren't feeling well.

58:42 WE WORKED AT HOME ALL DAY— GrammarRap
The GrammarRappers were very busy today.

SBS-TV Backstage Bulletin Board

TO: Production Crew

Sets and props for this segment:

Office
- clock
- potato chips
- coffee
- popcorn
- sandwich
- soda

Living Room
- coffee cups
- sofa

TV Studio
- paint roller
- wrench
- duster
- mop

TO: Cast Members

Key words in this segment:
- backache
- earache
- headache
- stomachache
- sore throat
- toothache
- cold

HOW DO YOU FEEL TODAY?

EDITING MIX-UP

The video editor made a mistake! Put each set of lines in the correct order.

a
___ I feel great!
_1_ Hi, Bob. How are you today?
___ I'm glad to hear that.

b
___ I'm happy to hear that.
___ I feel fine.
___ Hello, Bob. How are you today?

c
___ Thanks.
___ Okay.
___ That's good. Have a nice lunch, Bob.
___ Hi, Bob. How are you doing today?

d
___ Thanks, Nancy.
___ So-so.
___ Hi, Bob. How are you doing?
___ Oh. Well, have a good afternoon, Bob.

e
___ Not so good.
___ Oh. I'm sorry to hear that.
___ Hi, Bob. How are you?

f
___ I feel terrible.
___ See you tomorrow, Alan.
___ Bob, you don't look very well. Are you okay?
___ I'm sorry to hear that. Well, see you tomorrow, Bob.

PICTURE THIS!

Match the conversations above with the following scenes.

1 _c_

2 ____

3 ____

4 ____

5 ____

6 ____

THE NEXT LINE

Circle the correct response to each line in the scene.

1. I feel fine.
   a. I'm sorry to hear that.
   b. I'm happy to hear that. *(circled)*

2. How are you doing?
   a. So-so.
   b. I'm eating.

3. See you tomorrow.
   a. Hi!
   b. See you tomorrow.

4. Are you okay?
   a. Yes. I'm fine.
   b. That's good.

5. I feel terrible.
   a. I'm sorry to hear that.
   b. I'm happy to hear that.

6. Have a nice lunch.
   a. Not so good.
   b. Thanks.

MATCH THE SENTENCES!

Match the sentences that have the same meaning.

__d__ 1. I feel great!   a. Hello.
____ 2. How are you?   b. I'm happy to hear that.
____ 3. Hi!   c. How are you doing?
____ 4. I'm glad to hear that.   d. I feel fine.

FINISH THE SCRIPT!

Complete the following conversations and then practice them with a friend.

A. How are you doing?
B. _____
A. I'm happy to hear that.

A. How are you today?
B. _____
A. I'm sorry to hear that.

SEGMENT 24

56:45 SBS-TV ON LOCATION

PREVIEW

Help the cast rehearse important words in this segment.

| backache | earache | sore throat | toothache |
| cold | headache | stomachache | |

1. __headache__   2. _____   3. _____   4. _____

  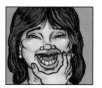

5. _____   6. _____   7. _____

SOUND CHECK

What's the matter?
I have . . .

1. a. **a headache**   2. a. a backache   3. a. a headache   4. a. a toothache
   b. an earache   b. a stomachache   b. a toothache   b. a backache

5. a. a headache   6. a. a sore throat   7. a. a cold
   b. an earache   b. a toothache   b. a sore throat

122

SEGMENT 24

57:07 SBS-TV ON LOCATION

SOUND CHECK

What did you do yesterday?

clean   cook   paint   plant   play   study   wait   wash   work

1. I ___worked___.
2. I _____.
3. I _____ my windows.
4. I _____ cards with my friends.
5. I _____ my house.
6. I _____.
7. I _____ my bathroom.
8. I _____ flowers.
9. I _____ for the plumber all day.

WHAT'S THE MATTER?

EDITING MIX-UP

The video editor made a mistake! Put the following lines in the correct order.

___ Not so good.

___ He has a backache.

_1_ How does David feel?

___ He played basketball all day.

___ Oh. I'm sorry to hear that.

___ What's the matter?

___ A backache? How did he get it?

FINISH THE SCRIPT!

These people don't feel well. Complete the following conversations and then practice them with a friend. In your conversations, use the words clean, paint, play, study, wait, wash, work.

A. How does Jennifer feel?

B. ....

A. What's the matter?

B. ....

A. ....? How did she get it?

B. .... all morning.

A. I'm sorry to hear that.

A. How does Frank feel?

B. ....

A. What's the matter?

B. ....

A. ....? How did he get it?

B. .... all afternoon.

A. I'm sorry to hear that.

A. How do you feel?

B. ....

A. What's the matter?

B. ....

A. ....? How did you get it?

B. .... all day.

A. I'm sorry to hear that.

SEGMENT 24

57:47 SBS-TV ON LOCATION

SOUND CHECK

How do you feel today?

eat   sing
drink sit

I have a ___stomachache___¹.

I ___ate___² cookies all afternoon.

I have a _____³.

I _____⁴ coffee all morning.

I have a terrible _____⁵.

I _____⁶ at my desk all day.

I have a very bad sore _____⁷.

I _____⁸ all day today.

THE NEXT LINE

Circle the correct response.

1. How do you feel?
   a. (Not so good.)
   b. Not really.

2. Are you feeling okay?
   a. Not so good.
   b. Not really.

3. How did you get it?
   a. I ate cookies all afternoon.
   b. Not so good.

4. I feel great!
   a. I'm glad to hear that.
   b. I'm sorry to hear that.

5. What did you do?
   a. Not so good.
   b. I sat at my desk all day today.

6. What's the matter?
   a. I have a headache.
   b. I played basketball.

7. I have a very bad sore throat.
   a. I'm sorry to hear that.
   b. How are you today?

8. Awful!
   a. What's the matter?
   b. How do you feel?

SEGMENT 24

125

WE WORKED AT HOME ALL DAY

FINISH THE RAP!

| clean | fix | paint | wash | floors | house | porch | sink |

What did you do today?

I _____washed_____¹ my

_____².

Your _____³?

Yes! I _____⁴ my

_____⁵ all day!

What did you do today?

I _____⁶ my

_____⁷.

Your _____⁸?

Yes! I _____⁹ my

_____¹⁰ all day!

What did you do today?

I _____¹¹ my

_____¹².

Your _____¹³?

Yes! I _____¹⁴ my

_____¹⁵ all day!

What did you do today?

I _____¹⁶ my

_____¹⁷.

Your _____¹⁸?

Yes! I _____¹⁹ my

_____²⁰ all day!

I _____²¹ my floors.

I _____²² my house.

I _____²³ my porch.

I _____²⁴ my sink.

We worked at home all day!

WRITE YOUR OWN RAP!

What did YOU do today? Write a GrammarRap about yourself.

What did you do today?

I .... my ....

Your .... ?

Yes! I .... my .... all day!

126

SEGMENT 24

WHAT'S THE RIGHT WORD?

1. Susan ~~clean~~ (cleaned) her house yesterday. Today she has a terrible ~~toothache~~ / backache.

2. When Tom is tired, he watches / ~~watched~~ TV.

3. Last night Francine ~~eats~~ / ate a lot. Today she has a very bad ~~headache~~ / stomachache.

4. Anthony played / ~~play~~ cards with his friends today. He played / ~~play~~ cards yesterday, too.

5. Martha waited / ~~wait~~ for her family to call her yesterday.

6. The president talked / ~~talks~~ on television last night and everyone listened / ~~listen~~.

7. I studied / ~~study~~ all day yesterday and I studied / ~~study~~ all day today, too. I'm really tired!

8. When I drink / ~~drank~~ a lot of coffee, I always get a bad headache / ~~backache~~.

9. What a busy day! We ~~cook~~ / cooked and cleaned / ~~clean~~ all morning!

10. I usually exercise / ~~exercised~~ in the morning. Yesterday I ~~exercise~~ / exercised in the afternoon.

WHAT DID THEY DO?

1. Mark doesn't usually play baseball, but he _____played_____ baseball all morning today.

2. Tanya never cleans her apartment, but she _____ it all day today.

3. I rarely eat ice cream, but I _____ ice cream yesterday afternoon.

4. Alice _____ her car yesterday. That's strange. She NEVER washes her car!

5. My children _____ dinner for me last night. My children don't usually cook dinner!

6. I have a backache. I _____ in a very uncomfortable chair all morning at work.

7. Of course you have a sore throat! You _____ all morning. I always get a sore throat when I sing.

8. I never drink coffee or tea, but I _____ a lot of coffee this weekend.

SEGMENT 24

SEGMENT 24 SCRIPT

55:15 HOW DO YOU FEEL TODAY?

(Bob is having a snack at the office.)
CO-WORKER 1: Hi, Bob. How are you today?
BOB: I feel great!
CO-WORKER 1: I'm glad to hear that.

(Bob is eating some more.)
CO-WORKER 2: Hello, Bob. How are you today?
BOB: I feel fine.
CO-WORKER 2: I'm happy to hear that.

(Bob is eating lunch.)
CO-WORKER 3: Hi, Bob. How are you doing today?
BOB: Okay.
CO-WORKER 3: That's good. Have a nice lunch, Bob.
BOB: Thanks.

(Bob is having an afternoon snack.)
CO-WORKER 4: Hi, Bob. How are you doing?
BOB: So-so.
CO-WORKER 4: Oh. Well, have a good afternoon, Bob.
BOB: Thanks, Nancy.

(Bob is still eating.)
CO-WORKER 5: Hi, Bob. How are you?
BOB: Not so good.
CO-WORKER 5: Oh. I'm sorry to hear that.

(Bob doesn't feel well.)
CO-WORKER 6: Bob, you don't look very well. Are you okay?
BOB: I feel terrible.
CO-WORKER 6: I'm sorry to hear that. Well, see you tomorrow, Bob.
BOB: See you tomorrow, Alan.

56:45 SBS-TV ON LOCATION

INTERVIEWER: What's the matter?
PERSON 1: I have a headache.
PERSON 2: I have a stomachache.
PERSON 3: I have a toothache.
PERSON 4: I have a backache.
PERSON 5: I have an earache.
PERSON 6: I have a sore throat.
PERSON 7: I have a cold.
(Person 7 sneezes.)
Excuse me.
INTERVIEWER: Bless you.

57:07 SBS-TV ON LOCATION

INTERVIEWER: What did you do yesterday?
PERSON 1: I worked.
PERSON 2: I cooked.
PERSON 3: I washed my windows.
PERSON 4: I played cards with my friends.
PERSON 5: I cleaned my house.
PERSON 6: I studied.
PERSON 7: I painted my bathroom.
PERSON 8: I planted flowers.
PERSON 9: I waited for the plumber all day.

57:28 WHAT'S THE MATTER?

MOTHER: How does David feel?
DAUGHTER: Not so good.
MOTHER: What's the matter?
DAUGHTER: He has a backache.
MOTHER: A backache? How did he get it?
DAUGHTER: He played basketball all day.
MOTHER: Oh. I'm sorry to hear that.

57:47 SBS-TV ON LOCATION

INTERVIEWER: How do you feel?
PERSON 1: Not so good.
INTERVIEWER: What's the matter?
PERSON 1: I have a stomachache.
INTERVIEWER: A stomachache? How did you get it?
PERSON 1: I ate cookies all afternoon.

INTERVIEWER: Are you feeling okay?
PERSON 2: Not really.
INTERVIEWER: What's the matter?
PERSON 2: I have a headache.
INTERVIEWER: A headache? That's too bad. How did you get it?
PERSON 2: I drank coffee all morning.

INTERVIEWER: How are you feeling?
PERSON 3: Not so great.
INTERVIEWER: What's wrong?
PERSON 3: I have a terrible backache.
INTERVIEWER: I'm sorry to hear that. What did you do?
PERSON 3: I sat at my desk all day today.

INTERVIEWER: How do you feel today?
PERSON 4: Awful!
INTERVIEWER: What's the matter?
PERSON 4: I have a very bad sore throat.
INTERVIEWER: I'm sorry to hear that. How did you get it?
PERSON 4: I sang all day today. And how do YOU feel today?
INTERVIEWER: I feel great!
PERSON 4: I'm glad to hear that.

58:42 WE WORKED AT HOME ALL DAY—GrammarRap

What did you do today?
  I washed my floors.
Your floors?
  Yes! I washed my floors all day!

What did you do today?
  I cleaned my house.
Your house?
  Yes! I cleaned my house all day!

What did you do today?
  I painted my porch.
Your porch?
  Yes! I painted my porch all day!

What did you do today?
  I fixed my sink.
Your sink?
  Yes! I fixed my sink all day!

I washed my floors.
  I cleaned my house.
I painted my porch.
  I fixed my sink.
We worked at home all day!

SEGMENT 24 SUMMARY

GRAMMAR

Past Tense

| I / He / She / It / We / You / They | worked yesterday. |

[t] I work**ed**. / I danc**ed**.

[d] I clean**ed** my house. / I play**ed** baseball.

[Id] I rest**ed**. / I shout**ed**.

Irregular Verbs

eat – ate
drink – drank
sing – sang
sit – sat

FUNCTIONS

Asking for and Reporting Information

How do you feel today?
How are you?
  I feel *great/fine/okay*.
  So-so.
  Not so good.
  I feel terrible.
  I don't feel very well today.

What's the matter?
What seems to be the problem?
  I have *a headache*.
  I have *a terrible headache*.

What did you do yesterday?
  I *worked*.

How *did he get a backache*?

Responding to Information

I'm glad to hear that.
I'm sorry to hear that.
That's too bad.

Checking Understanding

A backache?

Greeting People

How are you today?
How are you doing today?

Leave Taking

Have a nice *lunch*.
Have a good *afternoon*.

See you tomorrow.

SEGMENT 25

- Past Activities
- Past Tense:
  Questions
  Short Answers
  More Irregular Verbs

"They didn't do a lot today. They did a lot more yesterday . . . Side by Side."

PROGRAM LISTINGS

59:40 I BRUSHED MY TEETH
A mother greets her two children at the breakfast table before school.

1:00:46 SBS-TV ON LOCATION
People tell what they did over the weekend.

1:01:55 ON THE GO!
A busy family talks about their busy day.

1:02:40 SBS-TV ON LOCATION
People tell what they did yesterday.

1:04:01 PRESTO VITAMINS
Presto Vitamins changed the lives of everybody in this family.

1:04:35 PRESTO PET FOOD
A Presto Pet Food user tells why her pets are happy now.

1:05:28 PRESTO PRODUCTS
People tell why they like Presto Products.

SBS-TV Backstage Bulletin Board

TO: Production Crew
Sets and props for this segment:

Kitchen
 table
 juice
 cereal
 counter
 groceries
 fish tank

TO: Cast Members
Key words in this segment:

fish	buy	clean
floor wax	brush	dirty
hair	go	white
paint	rest	yellow
pet food	smile	comfortable
poem	visit	uncomfortable
shampoo	work	beautiful
sofa	before	ugly
teeth	energetic	dull
toothpaste	tired	shiny
vitamins	happy	
weekend	sad	

59:40 I BRUSHED MY TEETH

SOUND CHECK

you	your	did	brush	sleep	hair
		didn't	brushed		

A. Good morning, Jessica.

B. Hi, Mom.

A. ___Did___¹ you sleep well?

B. Yes, I _____².

A. Jessica?

B. Yes, Mom?

A. _____³ you _____⁴ your hair this morning?

B. Yes, I _____⁵.

A. It looks very nice.

B. Thanks, Mom.

A. Good morning, Jimmy.

B. Hi, Mom.

A. _____⁶ you _____⁷ well last night?

B. No, I _____⁸.

A. Oh. That's too bad. Jimmy?

B. Yes, Mom?

A. _____ _____ _____⁹ your hair this morning?

B. No, I _____¹⁰. I _____¹¹ my teeth.

A. Well, please brush _____ _____¹² before you go to school. Okay, honey?

B. Okay, Mom.

132

SEGMENT 25

SBS-TV ON LOCATION

WHAT DID THEY DO?

Saturday	Sunday
✔ work in my garden	

He ___worked in his garden___¹ all weekend.

Saturday	Sunday
✔ visit some old friends from college	
✔ talk	

She _____ ².

They _____ and _____ and _____³ about the good old days.

Saturday	Sunday
✔ play baseball with my kids	✔ stay home and rest

On Saturday, she _____ ⁴.

On Sunday, she _____ ⁵ in her yard.

Saturday
✔ prepare for Leonard's visit
✔ dust the furniture
✔ vacuum the rugs
✔ clean all the bathrooms
✔ wash the windows
✔ cook
✔ bake

On Saturday, he and his wife _____ ⁶.

They _____ ⁷.

They _____ ⁸.

They _____ ⁹.

They _____ ¹⁰.

They _____¹¹ a lot of food.

They _____¹² bread and cookies and a big apple pie.

SEGMENT 25

YES, NO, OR MAYBE?

		Yes	No	Maybe
1	Leonard and his family called on Sunday morning.	(Yes)	No	Maybe
2	They said, "We can visit you today."	Yes	No	Maybe
3	Leonard's children were sick.	Yes	No	Maybe
4	They enjoyed their visit.	Yes	No	Maybe
5	They didn't visit.	Yes	No	Maybe
6	They wanted to visit, but they didn't feel well.	Yes	No	Maybe
7	They're going to visit next month.	Yes	No	Maybe
8	Leonard's cousin rested on Sunday and Sheila ate the apple pie.	Yes	No	Maybe

CLOSE-UP

You're on Side by Side TV! Tell about your weekend.

1 Did you have a good weekend? ....

2 What did you do on Saturday? ....
....

3 What did you do on Sunday? ....
....

INTERVIEW

Interview two friends. What did they do on the weekend? Write their answers.

Name: ....   Name: ....

On Saturday, (he / she) ....   On Saturday (he / she) ....
....

On Sunday, (he / she) ....   On Sunday (he / she) ....
....

SEGMENT 25

1:01:55 ON THE GO!

WHERE DID THEY GO?

Put a check (✔) next to the places where they went.

___ to the bank	___ to the doctor	___ to the library
___ bowling	___ to the drug store	___ to the post office
___ to the dentist	___ to the laundromat	✔ to the supermarket

SOUND CHECK

| did | go | the | laundromat | bowling | to |
| didn't | went | this | library | | |

DAD: Hi, honey! Hi, Danny!

MOM: Hi, Bob!

DANNY: Hi, Dad!

MOM: What a busy day!

DAD: ___Did___¹ you ___² to the bank this afternoon?

MOM: No, we ___³. We ___⁴ to the supermarket.

DAD: Oh, I see. Danny, ___⁵ you ___⁶ to the doctor ___⁷ morning?

DANNY: No, I ___⁸. I went ___ ___⁹ dentist.

DAD: Oh, that's right.

DANNY: Dad, did you ___ ___¹⁰ today?

DAD: No, I didn't. ___¹¹ morning I ___¹² to the post office, then I ___¹³ to the drug store, then I went ___ ___¹⁴, and this afternoon I ___¹⁵ to the ___¹⁶.

DANNY: Boy, we sure are a busy family!

MOM: We're always on the ___¹⁷!

DANNY: I don't know where this day ___¹⁸!

SEGMENT 25

1:02:40 SBS-TV ON LOCATION

YES, NO, OR MAYBE?

		Yes	No	Maybe
1	He got up late.	(Yes)	No	Maybe
2	He got up early.	Yes	No	Maybe
3	He got up at 6:00 in the morning.	Yes	No	Maybe
4	He got up at noon.	Yes	No	Maybe

5	She had a big breakfast.	Yes	No	Maybe
6	She didn't have any breakfast.	Yes	No	Maybe
7	She had coffee and bread for breakfast.	Yes	No	Maybe

8	He went to work today.	Yes	No	Maybe
9	He took the subway to work.	Yes	No	Maybe
10	He missed the bus.	Yes	No	Maybe

11	She didn't do her homework.	Yes	No	Maybe
12	She did her exercises.	Yes	No	Maybe
13	She rested all night.	Yes	No	Maybe

14	She went shopping today.	Yes	No	Maybe
15	She bought bananas.	Yes	No	Maybe
16	She bought oranges.	Yes	No	Maybe

17	She read the newspaper.	Yes	No	Maybe
18	She read a novel.	Yes	No	Maybe
19	She read a book.	Yes	No	Maybe

20	He wrote a letter to his son.	Yes	No	Maybe
21	He wrote a letter to his daughter.	Yes	No	Maybe
22	He wrote a poem.	Yes	No	Maybe

SEGMENT 25

SOUND CHECK

•••••••••••••••••••••••••••••••• **Busy Day** ••••••••••••••••••••••••••••••••

I _____had_____¹ a busy day today.

I _____² and _____³ and _____⁴ all day.

I _____⁵ work late and _____⁶ the train.

I _____⁷ to walk home in the rain.

I didn't _____⁸ a thing to eat.

I _____⁹ the news and _____¹⁰ my feet.

I _____¹¹ a bath,

And then I _____¹².

I _____¹³ some milk,

And _____¹⁴ to bed.

WHAT'S MY LINE?

Did	buy	get up	have	take	write
	bought	got up	had	took	wrote

1. __Did__ you _____ the bus to school today?   No, I didn't. I _____ the subway.

2. _____ you _____ any milk at the supermarket?   I'm sorry. I didn't. But I _____ some juice.

3. _____ Jane _____ at 7:00 this morning?   No. She _____ late.

4. _____ you _____ a big dinner last night?   No. We _____ a small dinner. We didn't want to be late for the movie.

5. _____ you _____ to your grandmother?   Yes. I _____ her a very long letter.

PRESTO VITAMINS

SOUND CHECK

| bought | children | was | were | we're | tired | wife |

Before our family ____bought____¹ Presto Vitamins, we _____² always tired. I _____³ tired. My _____⁴ was tired. My _____⁵ were _____⁶, too. Now _____⁷ energetic, because WE _____⁸ Presto Vitamins. How about you?

PRESTO PET FOOD

EDITING MIX-UP

The video editor made a mistake! Put the following lines in the correct order.

____ My dog Homer was sad.

____ All my fish were sad, too.

____ Presto Dog Food for Homer.

____ Presto Cat Food for Friskie.

____ You know, before I bought Presto Pet Food, my pets were always sad.

____ And Presto Fish Food for all my little friends right here!

__1_ Here, my little friends. It's dinnertime!

____ My pets were all sad, but now they're happy, because I bought Presto Pet Food! How about you?

____ Presto Pet Food. Available at pet stores and supemarkets everywhere.

____ My cat Friskie was sad.

____ Now my pets are happy, because I give them Presto Pet Food.

1:05:28 PRESTO PRODUCTS

YES OR NO?

1	Her hair is always dirty.	Yes	**No**
2	Her hair is long.	Yes	No
3	Her hair isn't dirty now.	Yes	No

4	He never smiled.	Yes	No
5	He was embarrassed because his teeth were yellow.	Yes	No
6	He has a toothache.	Yes	No

7	Before she bought her Presto Sofa, she was never comfortable.	Yes	No
8	Her guests were uncomfortable.	Yes	No
9	Her guests never sit on her sofa.	Yes	No

10	He painted the rooms in his house.	Yes	No
11	He painted the outside of his house.	Yes	No
12	His house is ugly now.	Yes	No

13	His kitchen floor is really dull.	Yes	No
14	His kitchen floor was shiny before.	Yes	No
15	He drinks Presto Coffee.	Yes	No

EDITING MIX-UP 1

The video editor made a mistake! Put the following lines from the paint commercial in the correct order.

___ It was ugly inside and outside.

___ What a difference!

___ My house looks beautiful now . . . thanks to Presto.

_1_ Before I bought Presto Paint, my house was ugly.

___ But then I bought Presto Paint.

SEGMENT 25

EDITING MIX-UP 2

The video editor made another mistake! Put the following lines from the floor wax commercial in the correct order.

____ I mean, REALLY DULL.

____ It's just beautiful.

____ Then I bought Presto Floor Wax.

__1__ Let me tell you about my kitchen floor.

____ Every morning, I sit in my kitchen, drink a cup of Presto Coffee, and say, "Thank you Presto Floor Wax. You changed my life!"

____ It was dull.

____ My kitchen floor is shiny now.

FIX THE COMMERCIALS!

1. My pets are happy because I feed them Presto ~~Floor Wax~~. _____Pet Food_____

2. Now my teeth are white and I smile all the time thanks to Presto Paint. _____

3. My hair is clean thanks to Presto Coffee. _____

4. My house looks beautiful thanks to Presto Toothpaste. _____

5. Let me tell you about my kitchen floor. It was dull. Then I bought Presto Shampoo. _____

6. My guests are always comfortable when they sit on my Presto Vitamins. _____

7. Before our family bought Presto Pet Food, we were always tired. _____

WRITE YOUR OWN COMMERCIAL!

Before our family bought _____, we were always
                              (your product)
_____. I was _____. My (wife / husband)

was _____. My children were _____, too. Now we're

_____ because we bought _____.
                                      (your product)

How about you?

BAD WEATHER

What a surprise! It rained yesterday in Centerville and everybody in town had to change their plans. Look at the calendars below and tell what everybody DID and DIDN'T do.

1. He didn't go to the beach.
   He went to the movies.

2. _____
   _____

3. _____
   _____

4. _____
   _____

5. _____
   _____

6. _____
   _____

7. _____
   _____

8. _____
   _____

SEGMENT 25

TV CROSSWORD

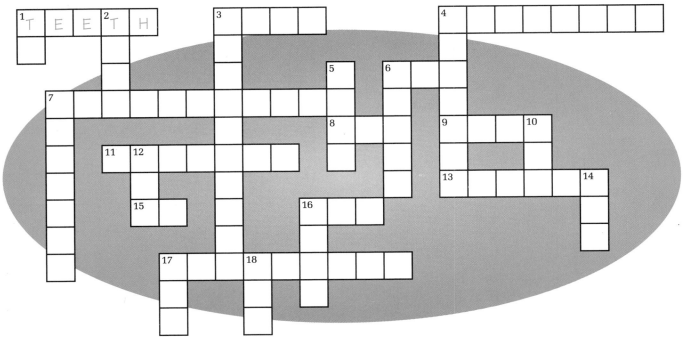

Across →

1. Before I bought Presto Toothpaste, my ____ were yellow.
3. We ____ the bus to work this morning.
4. I was tired, but now I'm energetic because I take ____ every day.
6. Our teacher ____ sick yesterday.
7. This Presto Sofa is very ____.
8. I'm happy, and my dog is happy, ____.
9. Did you go to the bank ____ afternoon?
11. Can I wash my hair with your ____?
13. They ____ their furniture. Now it's clean.
15. What did you ____ this weekend?
16. Are you happy or ____?
17. I ate a big ____ this morning.

Down ↓

1. We went ____ the supermarket.
2. They visit us, and we visit ____.
3. I brush my teeth with ____.
4. Yesterday, I ____ an old friend.
5. I have two ____: a cat and a dog.
6. I ____ a poem. Do you want to hear it?
7. Our house was very dirty, so we ____ it.
10. She ____ on the sofa and watched TV.
12. I ____ a bad headache last night.
14. ____ you brush your hair this morning?
16. I always sit on the ____ when I'm in the living room.
17. He didn't ____ bananas. He bought apples.
18. We ____ a lot of food last night.

SEGMENT 25 SCRIPT

59:40 I BRUSHED MY TEETH

MOTHER: Good morning, Jessica.
JESSICA: Hi, Mom.
MOTHER: Did you sleep well?
JESSICA: Yes, I did.
MOTHER: Jessica?
JESSICA: Yes, Mom?
MOTHER: Did you brush your hair this morning?
JESSICA: Yes, I did.
MOTHER: It looks very nice.
JESSICA: Thanks, Mom.
MOTHER: Good morning, Jimmy.
JIMMY: Hi, Mom.
MOTHER: Did you sleep well last night?
JIMMY: No, I didn't.
MOTHER: Oh. That's too bad. Jimmy?
JIMMY: Yes, Mom?
MOTHER: Did you brush your hair this morning?
JIMMY: No, I didn't. I brushed my teeth.
MOTHER: Well, please brush your hair before you go to school. Okay, honey?
JIMMY: Okay, Mom.

1:00:46 SBS-TV ON LOCATION

INTERVIEWER: Did you have a nice weekend?
PERSON 1: Yes, I did. I worked in my garden all weekend.

INTERVIEWER: Did you have a good weekend?
PERSON 2: Yes. I visited some old friends from college and we just talked and talked and talked about the good old days.

INTERVIEWER: What did you do this weekend?
PERSON 3: On Saturday I played baseball with my kids, and on Sunday I stayed home and rested in my yard.

INTERVIEWER: Did you have a nice weekend?
PERSON 4: No, I didn't.
INTERVIEWER: I'm sorry to hear that. What happened?
PERSON 4: Well, my cousin Leonard called on Saturday morning. He and his wife and three kids wanted to visit us on Sunday. So all day Saturday, my wife Sheila and I prepared for their visit. We dusted the furniture, we vacuumed the rugs, we cleaned all the bathrooms, we washed the windows, we cooked a lot of food, and we baked bread and cookies and a big apple pie.
INTERVIEWER: You really worked hard. Did your cousin and his family enjoy their visit?
PERSON 4: What visit? They didn't come!
INTERVIEWER: They didn't?
PERSON 4: No. They called Sunday morning and said, "Sorry. We can't visit you today."
INTERVIEWER: That's terrible! Did they say why?
PERSON 4: No, they didn't.
INTERVIEWER: Well, I'm sorry to hear about your weekend.
PERSON 4: Thanks.

1:01:55 ON THE GO!

DAD: Hi, honey! Hi, Danny!
MOM: Hi, Bob!
DANNY: Hi, Dad!
MOM: What a busy day!
DAD: Did you go to the bank this afternoon?

MOM: No, we didn't. We went to the supermarket.
DAD: Oh, I see. Danny, did you go to the doctor this morning?
DANNY: No, I didn't. I went to the dentist.
DAD: Oh, that's right.
DANNY: Dad, did you go bowling today?
DAD: No, I didn't. This morning I went to the post office, then I went to the drug store, then I went to the library, and this afternoon I went to the laundromat.
DANNY: Boy, we sure are a busy family!
MOM: We're always on the go!
DAD: I don't know where this day went!

1:02:40 SBS-TV ON LOCATION

INTERVIEWER: Did you get up early today?
PERSON 1: No, I didn't. I got up late.

INTERVIEWER: Did you have a big breakfast today?
PERSON 2: No, I didn't. I had a very small breakfast.

INTERVIEWER: Did you take the bus to work this morning?
PERSON 3: No, I didn't. I took the subway.

INTERVIEWER: Did you do your homework last night?
PERSON 4: No, I didn't. I did my exercises.

INTERVIEWER: Did you buy apples today?
PERSON 5: No, I didn't. I bought bananas. Do you want one?
INTERVIEWER: No, thanks.

INTERVIEWER: Did you read the newspaper today?
PERSON 6: No, I didn't. I read a novel.

INTERVIEWER: Did you write any letters today?
PERSON 7: No, I didn't, but I wrote a poem.
INTERVIEWER: A poem? Oh!
PERSON 7: Do you want to hear it?
INTERVIEWER: Hmm. I don't think we have time.
PERSON 7: It's a short one.
INTERVIEWER: Oh. Well, all right. Let's hear your poem.
PERSON 7: It's called "Busy Day."
I had a busy day today.
I wrote and read and worked all day.
I left work late and missed the train.
I had to walk home in the rain.
I didn't have a thing to eat.
I watched the news and rested my feet.
I took a bath,
And then I read.
I drank some milk,
And went to bed.
INTERVIEWER: That's a very nice poem!
PERSON 7: Thank you.

ANNOUNCER: Side by Side TV will return after these commercial messages.

1:04:01 PRESTO VITAMINS

HUSBAND: Before our family bought Presto Vitamins, we were always tired. I was tired. My wife was tired. My children were tired, too. Now we're energetic, because WE bought

144   SEGMENT 25

Presto Vitamins. How about you?

ANNOUNCER: Presto Vitamins. Available at drug stores and supermarkets everywhere.

1:04:35 PRESTO PET FOOD

ELDERLY LADY: *(To her fish.)*
Here, my little friends. It's dinnertime!
(To the viewer.)
You know, before I bought Presto Pet Food, my pets were always sad. My dog Homer was sad. My cat Friskie was sad. All my fish were sad, too.

Now my pets are happy, because I give them Presto Pet Food. Presto Dog Food for Homer. Presto Cat Food for Friskie. And Presto Fish Food for all my little friends right here!

My pets were all sad, but now they're happy, because I bought Presto Pet Food! How about you?

ANNOUNCER: Presto Pet Food. Available at pet stores and supermarkets everywhere.

1:05:28 PRESTO PRODUCTS

PERSON 1: Before I bought Presto Shampoo, my hair was always dirty. Now it's clean.

PERSON 2: Before I bought Presto Toothpaste, my teeth were always yellow. To tell the truth, I was very embarrassed. I never smiled. Now my teeth are white, and I smile all the time.

PERSON 3: Before I bought my Presto Sofa, I was always very uncomfortable when I sat in my living room, and I know my guests were uncomfortable, too. Now everybody's comfortable, thanks to my Presto Sofa!

PERSON 4: Before I bought Presto Paint, my house was ugly. It was ugly inside and outside. But then I bought Presto Paint. What a difference! My house looks beautiful now . . . thanks to Presto.

PERSON 5: Let me tell you about my kitchen floor. It was dull. I mean, REALLY DULL. Then I bought Presto Floor Wax. My kitchen floor is shiny now. It's just beautiful. Every morning, I sit in my kitchen, drink a cup of Presto Coffee, and say, "Thank you, Presto Floor Wax. You changed my life!"

ANNOUNCER: Consumers worldwide agree: If you aren't using Presto Products, you're living in the past. Presto . . . for the present!

SEGMENT 25 SUMMARY

GRAMMAR

Past Tense: Yes/No Questions

| Did | I / he / she / it / we / you / they | work? |

Short Answers

| Yes, | I / he / she / it / we / you / they | did. |

| No, | I / he / she / it / we / you / they | didn't. |

Past Tense: WH-Questions

| What did | I / he / she / it / we / you / they | do? |

To Be: Past Tense

| I / He / She / It | was | happy. |
| We / You / They | were | |

Irregular Verbs

buy – bought
drink – drank
do – did
get – got
have – had
leave – left
read – read
take – took
write – wrote

Time Expressions

| Did you study English | yesterday? / yesterday morning? / yesterday afternoon? / yesterday evening? / last night? |

FUNCTIONS

Asking for and Reporting Information

Did you *go to the bank this afternoon?*
Yes, I did.
No I didn't.

You know, . . .

Let me tell you about *my kitchen floor.*

Complimenting

It looks very nice.

Greeting People

Hi.

SEGMENT 26

- Past Activities
- To Be: Past Tense

"I wasn't there. I didn't go. You didn't care. I didn't know.

But now we're . . . Side by Side."

PROGRAM LISTINGS

1:07:00 WERE YOU AT THE BALLGAME LAST NIGHT?
A young boy is upset because he thinks nobody in his family saw him play in a baseball game.

1:08:11 DID YOU . . . ?
Some people did. Some people didn't.

1:08:40 I REMEMBER
An elderly couple reminisces about the time they were young.

SBS-TV Backstage Bulletin Board

TO: Production Crew
Sets and props for this segment:

Bedroom
  bed
  chair
  computer game

Kitchen
  table
  chairs
  cups
  plates
  orange juice
  pancakes
  cereal
  toast

Porch
  rocking chairs
  knitting needles
  newspaper

TO: Cast Members
Key words in this segment:

was   did
wasn't   didn't
were
weren't

WERE YOU AT THE BALLGAME LAST NIGHT?

WHERE WAS EVERYBODY?

was	were	meeting	computer class	ballgame
		birthday party	movies	

1. Tommy's brother __was__ at the _____.
2. Tommy's father _____ at a _____.
3. Tommy's mother _____ at her _____.
4. Tommy's sisters _____ at a _____.
5. Tommy's grandparents _____ at the _____.

SOUND CHECK

was	wasn't	were	weren't

I can't believe it! Mom __wasn't__¹ there, Dad _____² there, my sisters _____³ there, my brother _____⁴ there. Nobody _____⁵ at my ballgame.

That's not true, Tommy. Somebody _____⁶ at your ballgame. Grandma and Grandpa _____⁷ there.

THE NEXT LINE

Circle the correct response.

1. Can I ask you a question?
   a. Morning, Tommy.
   b.) Sure, Tommy.

2. Were you at the ballgame last night?
   a. Yes, they were.
   b. No, I wasn't.

3. I was at the movies.
   a. Where were you?
   b. Oh, I see.

4. Grandma and Grandpa were there.
   a. They were?
   b. How about Grandma and Grandpa?

5. They were?
   a. Yes, I was.
   b. Yes, they were.

6. Sorry I wasn't there.
   a. Oh, I see.
   b. That's okay.

SEGMENT 26

DID YOU...?

SOUND CHECK

| did | didn't | was | wasn't | were | weren't |

A. ___Did___¹ you sleep well last night?
B. Yes, I _____². I _____³ tired.

A. _____⁴ Roger sleep well last night?
B. No, he _____⁵. He _____⁶ tired.

A. _____⁷ you have a big breakfast today?
B. Yes, we _____⁸. We _____⁹ very hungry.

A. _____¹⁰ Marcela and Carlos have a big breakfast today?
B. No, they _____¹¹. They _____¹² hungry.

INTERVIEW

Interview two friends. Ask about things they did and write their answers in your reporter's notebook.

	Friend 1	Friend 2
1 Did you sleep well last night?		
2 Did you have a big breakfast today?		
3 Did you have a big dinner last night?		
4 Did you _____ ?		
5 Did you _____ ?		

SEGMENT 26

I REMEMBER

ON CAMERA

Fill in the missing words and then, with a friend, role-play the scene with Gertrude and Harold.

| was | were | did | go | eat | start | talk |
| wasn't | weren't | didn't | sit | have | rain | look |

HAROLD: Gertrude?

GERTRUDE: Yes, Harold?

HAROLD: Do you remember when we _____were_____ [1] young?

GERTRUDE: Of course I do. I remember like it _____ [2] yesterday.

HAROLD: I remember the day we met. You _____ [3] fourteen. I _____ [4] fifteen.

GERTRUDE: Actually, I _____ [5] sixteen, and you _____ [6] seventeen.

HAROLD: I _____ [7] ?

GERTRUDE: Yes, you _____ [8].

HAROLD: You _____ [9] beautiful.

GERTRUDE: Oh no, I _____ [10].

HAROLD: Yes, you _____ [11], Gertrude. You _____ [12] very beautiful.

GERTRUDE: You _____ [13] handsome.

HAROLD: Me, handsome? No, I _____ [14]. I _____ [15] nothing special.

GERTRUDE: That's not true. You _____ [16] very handsome, and very special. I remember I _____ [17] shy.

HAROLD: You _____ [18] shy at all. You _____ [19] very outgoing and very popular.

GERTRUDE: Oh. That's right. I _____ [20] very popular. And you _____ [21] very popular, too. You _____ [22] always very busy.

HAROLD: I _____ [23]. I _____ [24] always with my friends.

GERTRUDE: Do you remember our first date?

SEGMENT 26

HAROLD: Of course. We _____25_____ to the zoo.

GERTRUDE: No, we _____26_____. We _____27_____ to the park.

HAROLD: We _____28_____ _____29_____ to the park, Gertrude. We _____30_____ to the zoo. I remember.

GERTRUDE: Harold, we _____31_____ to the park. We _____32_____ on a bench in the park, and we _____33_____ for hours and hours.

HAROLD: Gertrude, we _____34_____ _____35_____ to the park, and we _____36_____ _____37_____ on a bench. We _____38_____ to the zoo and _____39_____ at the animals

GERTRUDE: You're wrong, Harold. I remember. I remember like it _____40_____ yesterday.

HAROLD: You know what? The zoo _____41_____ in the park!

GERTRUDE: That's right. It _____42_____. We _____43_____ to the zoo in the park.

HAROLD: We _____44_____. And we _____45_____ on a bench, and we _____46_____ for hours and hours.

GERTRUDE: We _____47_____ some ice cream.

HAROLD: That's right. And we _____48_____ some popcorn.

GERTRUDE: That's right. We _____49_____.

HAROLD: It _____50_____ a beautiful day.

GERTRUDE: Actually, it _____51_____ to _____52_____.

HAROLD: It _____53_____ _____54_____. It _____55_____ sunny!

GERTRUDE: No, it _____56_____, Harold. It _____57_____ sunny. It _____58_____.

HAROLD: It _____59_____ _____60_____, Gertrude.

GERTRUDE: Yes, it _____61_____.

HAROLD: Well, maybe you're right.

GERTRUDE: It doesn't matter. It _____62_____ a very special day.

HAROLD: You're right. It _____63_____. I remember it like it _____64_____ yesterday.

SEGMENT 26

WHAT'S THE RESPONSE?

Choose the best answer.

__j__ 1. Did you and your husband watch TV last night?
____ 2. Was your brother late for class?
____ 3. Did Mrs. Cummings go to the meeting at her son's school?
____ 4. Did my brother call me last night?
____ 5. Were you popular when you were a young boy?
____ 6. Were you and your wife on time for the concert?
____ 7. Was your sister at home yesterday?
____ 8. Did your friends come to your baseball game?
____ 9. Was I a quiet baby?
____ 10. This food is delicious. Did you cook it?

a. No, you weren't.
b. No, they didn't.
c. Yes, we were.
d. Yes, I was.
e. No, I didn't.
f. No, he wasn't.
g. Yes, he did.
h. Yes, she did.
i. Yes, she was.
j. Yes, we did.

WHAT'S THE WORD?

| was | wasn't | were | weren't | did | didn't |

1. I ___was___ tired today because I studied all night.
2. _____ you have a good time at the picnic?
3. We didn't eat a very large lunch because we _____ very hungry.
4. We met when we _____ eighteen years old.
5. It _____ a beautiful day yesterday. It _____ rain, and it _____ cloudy.
6. I'm sorry you _____ at the party. Everybody had a wonderful time.
7. I _____ sleep well last night because I _____ very upset.
8. Sally _____ late for work because she _____ at the train station on time.
9. I _____ do my exercises this morning because I _____ sick.
10. Where _____ you? We needed you! Why _____ you there? Why _____ you call? We _____ all very concerned. _____ you okay, or _____ there a problem?

MATCH THE LINES!

_f_ 1. Fred didn't sleep very well because he ____.   a. weren't hungry
____ 2. We had a big breakfast today because we ____.   b. was tired
____ 3. My boss was upset this morning because I ____.   c. wasn't late
____ 4. I drank all the juice because I ____.   d. were hungry
____ 5. My children didn't eat a big dinner because they ____.   e. wasn't thirsty
____ 6. Billy didn't finish his milk because he ____.   f. wasn't tired
____ 7. I didn't miss the plane because I ____.   g. was thirsty
____ 8. I left the party early because I ____.   h. was late

SCRAMBLED SOUND TRACK

1. meeting | father | . | at | Tommy's | was | a

   Tommy's father was at a meeting.

2. your | you | class | Were | ? | computer | at

   _____

3. rain | sunny | It | It | . | . | was | didn't

   _____

4. ballgame | grandparents | the | at | . | were | Tommy's

   _____

5. dinner | hungry | we | We | big | were | . | had | because | a

   _____

6. the | the | at | zoo | went | looked | and | . | animals | to | They

   _____

7. yesterday | very | work | to | . | I | I | because | go | was | didn't | sick

   _____

8. ice cream | ate | eat | We | We | popcorn | . | . | didn't

   _____

9. and | park | , | and | They | on | the | they | a | talked | bench | hours | . | in | for | hours | sat

   _____

SEGMENT 26

SEGMENT 26 SCRIPT

1:07:00 WERE YOU AT THE BALLGAME LAST NIGHT?

TOMMY: Hi, Jeff.
JEFF: Morning, Tommy.
TOMMY: Jeff, can I ask you a question?
JEFF: Sure, Tommy.
TOMMY: Were you at the ballgame last night?
JEFF: No, I wasn't. I was at the movies.
TOMMY: Oh, I see. Was DAD at my baseball game?
JEFF: No, he wasn't. He was at a meeting.
TOMMY: How about Mom? Was SHE there?
JEFF: No, Tommy, she wasn't. She was at her computer class.
TOMMY: How about Katie and Melissa? Were THEY at my game?
JEFF: No, they weren't. They were at Jennifer Henderson's birthday party.
TOMMY: I can't believe it! Mom wasn't there, Dad wasn't there, my sisters weren't there, my brother wasn't there. Nobody was at my ballgame.
JEFF: That's not true, Tommy.
TOMMY: Huh?
JEFF: Somebody WAS at your ballgame.
TOMMY: Who?
JEFF: Grandma and Grandpa were there.
TOMMY: They WERE?
JEFF: Yes, they were. And you know what they said?
TOMMY: No. What?
JEFF: You were terrific.
TOMMY: I was terrific?
JEFF: They said you were GREAT!
TOMMY: Yes!
JEFF: Hey, Tommy?
TOMMY: Yeah?
JEFF: Sorry I wasn't there.
TOMMY: That's okay. Next time.

1:08:11 DID YOU...?

INTERVIEWER: Did you sleep well last night?
PERSON 1: Yes, I did. I was tired.

INTERVIEWER: Did Roger sleep well last night?
PERSON 2: No, he didn't. He wasn't tired.

INTERVIEWER: Did you have a big breakfast today?
PERSON 3: Yes, we did. We were very hungry.

INTERVIEWER: Did Marcela and Carlos have a big breakfast today?
PERSON 4: No, they didn't. They weren't hungry.

1:08:40 I REMEMBER

HAROLD: Gertrude?
GERTRUDE: Yes, Harold?
HAROLD: Do you remember when we were young?
GERTRUDE: Of course I do. I remember like it was yesterday.
HAROLD: I remember the day we met. You were fourteen. I was fifteen.
GERTRUDE: Actually, I was sixteen, and you were seventeen.
HAROLD: I was?
GERTRUDE: Yes, you were.
HAROLD: You were beautiful.
GERTRUDE: Oh no, I wasn't.
HAROLD: Yes, you were, Gertrude. You were very beautiful.
GERTRUDE You were handsome.
HAROLD: Me, handsome? No, I wasn't. I was nothing special.
GERTRUDE: That's not true. You were very handsome, and very special. I remember I was shy.
HAROLD: You weren't shy at all. You were very outgoing and very popular.
GERTRUDE: Oh. That's right. I WAS very popular. And you were very popular, too. You were always very busy.
HAROLD: I was. I was always with my friends.
GERTRUDE: Do you remember our first date?
HAROLD: Of course. We went to the zoo.
GERTRUDE: No, we didn't. We went to the park.
HAROLD: We didn't go to the park, Gertrude. We went to the zoo. I remember.
GERTRUDE: Harold, we went to the park. We sat on a bench in the park, and we talked for hours and hours.
HAROLD: Gertrude, we didn't go to the park, and we didn't sit on a bench. We went to the zoo and looked at the animals.
GERTRUDE: You're wrong, Harold. I remember. I remember like it was yesterday.
HAROLD: You know what? The zoo was in the park!
GERTRUDE: That's right. It was. We went to the zoo in the park.
HAROLD: We did. And we sat on a bench, and we talked for hours and hours.
GERTRUDE: We ate some ice cream.
HAROLD: That's right. And we had some popcorn.
GERTRUDE: That's right. We did.
HAROLD: It was a beautiful day.
GERTRUDE: Actually, it started to rain.
HAROLD: It didn't rain. It was sunny!
GERTRUDE: No, it wasn't, Harold. It wasn't sunny. It rained.
HAROLD: It didn't rain, Gertrude.
GERTRUDE: Yes, it did.
HAROLD: Well, maybe you're right.
GERTRUDE: It doesn't matter. It was a very special day.
HAROLD: You're right. It was. I remember it like it was yesterday.

SEGMENT 26 SUMMARY

GRAMMAR

To Be: Past Tense

I / He / She / It	**was**	happy.
We / You / They	**were**	

I / He / She / It	**wasn't**	tired.
We / You / They	**weren't**	

Was	I / he / she / it	late?
Were	we / you / they	

Yes,	I / he / she / it	**was.**
	we / you / they	**were.**

No,	I / he / she / it	**wasn't.**
	we / you / they	**weren't.**

FUNCTIONS

Asking for and Reporting Information

How about you?

Were you *at the ballgame last night?*
  No, I wasn't. I was *at the movies.*

Apologizing

Sorry *I wasn't there.*

Remembering

Do you remember *when we were young?*

Initiating a Topic

Can I ask you a question?

Attracting Someone's Attention

Gertrude?
Jeff, can I ask you a question?

Correcting

That's not true.

Actually, . . .

Agreeing

That's right.

It was.
We did.

Disagreeing

That's not true.

ANSWER KEY

SEGMENT 14

Page 2
SOUND CHECK 1
1. a
2. b
3. a
4. a
5. a
6. b
7. a

Pages 2–3
SOUND CHECK 2
1. cook
2. cooks
3. does
4. cook
5. cooks
6. Does
7. cook
8. does
9. cook
10. does

Page 4
SOUND CHECK 3
1. Does
2. doesn't
3. does
4. cook
5. cooks
6. Does
7. cook
8. doesn't
9. does
10. cook
11. cooks

Page 5
SOUND CHECK
1. Do
2. go
3. do
4. like
5. Do
6. go
7. don't
8. don't like
9. do
10. like
11. like
12. do
13. go
14. don't
15. doesn't cook

Page 7
SCRAMBLED SOUND TRACK
1. Do you go to Stanley's Restaurant on Tuesday?
2. What kind of food do you like?
3. Stanley doesn't cook French food.
4. Does Stanley cook Chinese food on Friday?
5. I don't go to Stanley's Restaurant on Sunday because I don't like American food.

Page 8
EDITING MIX-UP
2
10
5
11
7
13
4
12
9
8
3
6
1

STANLEY'S FAVORITE CUSTOMERS
1. She goes to Stanley's Restaurant on Wednesday.
   She speaks Chinese, eats Chinese food, drinks Chinese wine, and listens to Chinese music.
2. He goes to Stanley's Restaurant on Saturday.
   He speaks Spanish, eats Mexican food, drinks Mexican wine, and listens to Mexican music.
3. They go to Stanley's Restaurant on Tuesday.
   They speak Greek, eat Greek food, drink Greek wine, and listen to Greek music.

Page 9
WHAT'S MY LINE?
1. like, likes
2. works, works
3. listen, listen
4. study, studies
5. go, goes
6. speak, speaks, speaks, speak
7. cooks, cooks, cooks

DO THEY OR DON'T THEY?
1. do
2. does
3. doesn't
4. do
5. does
6. don't
7. don't
8. do
9. doesn't
10. don't

SEGMENT 15

Pages 14–15
SCRIPT CHECK
1. d
2. e
3. b
4. f
5. a
6. c
7. b
8. c
9. a
10. b
11. e
12. a
13. d
14. c
15. e
16. c
17. a
18. b
19. d
20. b
21. e
22. f
23. c
24. d
25. a

Page 16
CAN YOU PREDICT?
1. a
2. a
3. b
4. b
5. a
6. b

SOUND CHECK
1. a
2. b
3. b
4. a
5. b
6. a

Page 17
VIDEO EDITOR
1. a, c
2. b, d
3. b, d

CAN YOU PREDICT?
1. b
2. b
3. a
4. b
5. a

WHAT'S MY LINE?
1. game shows
2. news programs

Page 18
CAN YOU PREDICT?
1. b
2. a
3. a
4. a
5. b

SOUND CHECK
1. b
2. a
3. b
4. b
5. b

Page 19

CAN YOU PREDICT?
1. b
2. a
3. a
4. b
5. b
6. a

Page 20

GUESTS AND HOST
1. b
2. a
3. b
4. b

EDITING MIX-UP 1
5
7
8
2
4
3
1
6

Page 21

WHAT'S THE LINE?
1. a
2. a
3. b
4. a
5. b
6. a

SOUND CHECK 1
1. music
2. like
3. kind
4. you
5. what
6. of
7. like
8. your
9. do
10. like
11. jazz
12. does
13. like
14. doesn't
15. likes
16. don't
17. do
18. she
19. does
20. I

Page 22

SOUND CHECK 2
1. sport
2. do
3. you
4. like
5. which
6. does
7. like
8. your
9. is
10. Dave's
11. Baseball
12. What's
13. your
14. Football
15. don't
16. like
17. do
18. LOVE
19. don't
20. do
21. play
22. Saturday

EDITING MIX-UP 2
2
8
5
7
1
3
6
4

Page 23

SCRAMBLED WORDS
1. favorite
2. comedies
3. novels
4. golf
5. poetry
6. science fiction
7. classical music
8. adventure movies
9. performer
10. cartoons
11. short story

SEGMENT 16

Page 30

SOUND CHECK
1. you
2. me
3. him
4. her
5. them
6. us
7. it
8. you
9. her
10. me

Page 31

SOUND CHECK
1. b, d
2. b, c
3. a, d
4. a
5. b
6. always
7. usually
8. sometimes
9. rarely
10. never

Page 32

SCENE CHECK
1. is
2. rarely
3. usually

SOUND CHECK 1
1. never
2. sometimes
3. always

Pages 32–33

SOUND CHECK 2
1. wash
2. washes
3. do
4. washes
5. wash
6. wash
7. washes
8. do
9. cleans
10. cook

Page 33

SOUND CHECK 3
1. rarely
2. never
3. sometimes
4. usually
5. never, always

Page 34

FINISH THE RAP!
1. always
2. usually
3. sometimes
4. never
5. never
6. always
7. usually
8. sometimes
9. rarely
10. never
11. never

Page 35

SCRAMBLED WORDS
1. always
2. rarely
3. sometimes
4. usually
5. never

WHAT'S MY LINE?
1. wash, washes
2. clean
3. studies
4. watch, watches
5. call, calls
6. get, get
7. sings, sing
8. fixes, fix
9. washes, wash, wash

SEGMENT 17

Page 40

WHAT DO THEY HAVE?
1. a, c
2. a
3. a, b
4. b, c
5. a, b
6. b
7. b

Page 41

SOUND EFFECTS MIX-UP
1. f
2. a
3. c
4. d
5. e
6. b

WHAT'S MY LINE?
1. Do, have I have
2. Does, have she has
3. Do, have we have
4. Does, have it has
5. Do, have they have
6. Does, have he has

Page 42

SOUND CHECK

1. has
2. brown
3. have
4. has
5. long
6. I'm
7. short
8. don't
9. have
10. has
11. apartment
12. has
13. have
14. dog
15. have
16. have
17. has
18. guitar
19. has
20. have
21. bicycle
22. has
23. color
24. has
25. have
26. two
27. sisters
28. friends

Page 43

WHICH SISTER?

1. brown
2. blue
3. short
4. long
5. short
6. tall
7. an apartment
8. a house
9. dog
10. cat
11. guitar
12. piano
13. bicycle
14. car
15. color
16. black-and-white
17. just one or two
18. a lot of

Page 44

TV CROSSWORD

See page 165.

SEGMENT 18

Page 48

SOUND CHECK

1. crying
2. I'm crying
3. I
4. cry
5. I'm

SOUND CHECK

1. shivering
2. We're shivering
3. We
4. shiver
5. we're

TELL ME WHY!

1. He's
2. He
3. blushes
4. he's
5. She's
6. she's
7. She
8. sings
9. she's
10. They're
11. they're
12. They
13. dance
14. they're

Page 49

SOUND CHECK

1. yawning
2. yawning
3. yawning
4. I'm
5. I
6. yawn
7. I'm
8. asking
9. ask
10. leave
11. shouting
12. shouting
13. I'm
14. I
15. shout
16. I'm
17. leave
18. leaving

Page 50

WHICH CAPTION?

1. nervous
2. sad
3. happy
4. tired
5. sick
6. cold
7. hot
8. hungry
9. thirsty
10. angry
11. embarrassed

WHAT'S MY LINE?

1. hot
2. hungry
3. angry
4. cold
5. happy
6. thirsty
7. nervous

Page 51

SCRAMBLED SOUND TRACK

a. When I'm nervous, I giggle.
b. When I'm angry, I shout.
c. When I'm happy, I smile.
d. When I'm nervous, I bite my nails.
e. I never get angry.
f. When I'm angry, my face turns red.
g. When I'm happy, I sing.
h. When I'm nervous, I perspire.
i. When I'm happy, I whistle.

MATCH THE LINES

1. d
2. h
3. a
4. b
5. f
6. e
7. i
8. c
9. g

Page 52

FINISH THE RAP!

1. smile
2. frown
3. blush
4. shout
5. smiling
6. frowning
7. blushing
8. shouting
9. happy
10. sad
11. embarrassed
12. mad
13. smile
14. frown
15. blush
16. shout

Page 53

WHAT'S MY LINE?

1. I'm
2. bite
3. smiling
4. smile
5. turns
6. is turning
7. I'm relaxing
8. sleep
9. I
10. cry
11. giggling
12. I'm
13. asking
14. ask

SEGMENT 19

Page 58

EDITING MIX-UP

3
5
1
6
7
4
2

SOUND CHECK

1. are
2. doing
3. I'm
4. washing
5. That's
6. Do
7. wash
8. I
9. wash
10. I'm
11. washing
12. TODAY
13. are
14. doing
15. is
16. I'm
17. hear

Page 59

EDITING MIX-UP

5
7
1
10
8
3
11
2
4
9
6

WHAT'S THE LINE?

1. washing, wash
2. drinking, drink, drink
3. brushing, brush
4. feed, feeding

Page 61

FINISH THE RAP!

1. doing
2. working
3. late
4. doing
5. works
6. What's
7. He's
8. Cooking
9. he
10. that
11. He
12. cooks
13. on
14. doing
15. He's
16. his
17. Bathing
18. his
19. Why's
20. It's
21. always
22. his

Page 62

WRONG LINE

1. study
   studying
   at
   in
2. She
   Her
   dishes
3. usually
4. Are
   Does
   eats
   eating
5. watches
   watching
   today
6. shines
   shining
7. you
   your
8. they
   you

SCRAMBLED SOUND TRACK

1. He always cleans his apartment on Friday.
2. Do you usually wash the dishes in the bathtub?
3. She's walking to school because her bicycle is broken.
4. I never work late, but I'm working late today.
5. Does he usually cook spaghetti on Wednesday?
6. Why are you washing your dishes with "Ordinary Soap?"

SEGMENT 20

Page 66

SOUND CHECK

1. can
2. can't, can't
3. can
4. can't, can't

WHAT CAN THEY DO?

1. a
2. b
3. b
4. a
5. a
6. b

Page 67

SOUND CHECK

1. Can 2. can't 3. can

INFORMATION CHECK

1. b, c
2. a, c
3. b, c
4. c
5. a, c
6. b, c

Page 68

SCRAMBLED SOUND TRACK

A. Can Jack fix cars?
B. Of course he can. He fixes cars every day. He's a mechanic!

SCRIPT CHECK

1. baker
2. truck driver
3. teacher
4. chef
5. painter
6. writer
7. bus driver
8. dancer
9. secretary

Page 69

SCENE CHECK

1. can
2. can't
3. can
4. can't
5. can
6. can't
7. can't
8. can't
9. can't
10. can't
11. can't
12. can't
13. can

EDITING MIX-UP

1. 2 / 1
2. 2 / 1
3. 1 / 2
4. 2 / 1
5. 1 / 2
6. 2 / 1
7. 2 / 1
8. 2 / 1

Page 70

GOOD NEWS OR BAD NEWS?

1. Good news
2. Bad news
3. Good news
4. Good news
5. Bad news
6. Bad news
7. Good news
8. Bad news
9. Bad news
10. Good news
11. Good news

TO BE OR NOT TO BE AN ACTOR!

I can't bake, I can't drive a bus, I can't cook, I can't type, I can't teach, I can't paint, I can't drive a truck, I can't dance, I can't write.

Page 71

CAN THEY OR CAN'T THEY?

1. can, secretary
2. can't
3. can, baker
4. teacher, can't
5. mechanic, can
6. truck driver, can
7. chef, can
8. painter, can't

SEGMENT 21

Pages 76–77

SOUND CHECK

1. can't go
2. Can
3. I can't
4. work
5. go
6. have to clean
7. can your
8. go
9. they can't
10. do
11. their
12. party
13. can
14. you
15. have to go
16. can't
17. has to go
18. can't
19. go to
20. has to
21. to
22. doctor
23. I'm
24. can't
25. have to do
26. can you
27. my
28. Me
29. I can't
30. have to
31. work
32. I

Page 78

INFORMATION CHECK

✔	
✔	✔

EDITING MIX-UP 1

4
2
5
1
6
3
7

EDITING MIX-UP 2

5
3
7
2
6
1
4

Page 79

EDITING MIX-UP 3
3
1
7
5
4
6
2

WHAT'S JULIE SAYING?
1. c
2. b
3. a
4. b
5. a
6. b

Page 81

FINISH THE RAP!
1. can't
2. talk
3. We can't
4. now
5. have to
6. can't stop
7. I have to
8. can't
9. catch
10. We can't
11. We have to
12. stop
13. catch
14. now
15. We
16. can't
17. catch

Page 82

CAST PARTY

    I'm   **I'm**
~~I~~ very sorry, but ~~I~~ afraid John and I
can't   **your**   **on**
~~cant~~ ~~to~~ come to ~~you're~~ party ~~at~~
          has   **visit**
Friday. John ~~have~~ to ~~visits~~ his

parents in New York, and I have to
work
~~working~~ late.
       for
   Thanks ~~because~~ inviting us.

Page 83

THE NEXT LINE
1. a
2. b
3. a
4. a
5. b
6. a

WRONG LINE
1. c
2. b
3. d
4. b
5. b
6. d

SEGMENT 22

Page 88

SOUND CHECK
1. I'm going to work
2. He's going to fix
3. I'm going to fix
4. she's going to work
5. It's going to be
6. We're going to clean
7. I'm going to clean
8. she's going to clean
9. You're going to clean
10. I'm going to clean
11. He's going to clean
12. going to
13. They're going to clean

Page 89

WHAT'S HAPPENING?
1. b
2. d
3. e
4. h
5. f
6. g
7. c
8. a

EDITING MIX-UP
1. 2 / 1
2. 1 / 2
3. 2 / 1
4. 1 / 2
5. 2 / 1
6. 2 / 1

Page 90

WHAT'S HAPPENING?
1. b
2. b
3. a
4. b
5. b

THE NEXT LINE
1. a
2. a
3. b
4. a
5. b
6. b

Page 91

WHOSE LINE?
a. Helen
b. Howard
c. Helen
d. Howard
e. Howard

PICTURE THIS!
1. a
2. c
3. b
4. e
5. d

Page 92

WHOSE LINE?
1. Theodore
2. Theodore
3. Lance
4. Theodore
5. Lance
6. Lance
7. Theodore
8. Theodore
9. Theodore
10. Lance
11. Theodore
12. Lance

EDITING MIX-UP
3
6
1
4
8
2
5
9
7

Page 93

EDITING MIX-UP
2
6
4
3
1
5

SOUND CHECK
1. right now
2. immediately
3. At once
4. right away

Pages 94–95

LISTENING CHALLENGE!
1. True
2. True
3. False
4. True
5. True
6. True
7. True
8. False
9. True
10. False
11. True
12. False
13. True
14. False
15. True
16. False
17. True
18. True
19. False
20. False
21. True
22. False
23. True
24. True
25. False
26. True
27. True
28. False
29. True
30. True
31. True
32. True
33. False
34. True
35. True
36. False

Page 96

SCENE CHECK
1. She's
2. He's
3. March, brother
4. April, sister
5. She's
6. He's
7. He's, July
8. August, always
9. He's
10. She's
11. November
12. isn't

SCENE REVIEW
1. September, October
2. January, February
3. March, April, July, August

Page 97
PICTURE THIS!
1. b
2. e
3. f
4. a
5. d
6. c

Page 98
WHAT'S THE QUESTION?
1. When are you going to
2. Where are you going to
3. Why are you going to
4. Who are you going to
5. What are you going to

WHAT'S MY LINE?
1. going
2. have, evening
3. is, weekend
4. are
5. now
6. are you going to
7. going to
8. am
9. once
10. going to get

SEGMENT 23

Page 106
YES OR NO?
1. Yes
2. No
3. Yes
4. Yes
5. Yes
6. No
7. Yes
8. Yes
9. Yes
10. Yes
11. No

A GOOD DAY FOR THE BEACH
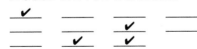

Page 107
WEATHER CHALLENGE!
1. cool
2. warm
3. hot
4. snow
5. clear
6. sunny
7. hot
8. warm
9. foggy
10. warm
11. hot
12. rain
13. drizzle
14. warm
15. sunny
16. hot
17. clear
18. cool
19. snow

Page 108
EDITING MIX-UP
4
3
1
6
5
2

SOUND CHECK
1. What
2. it
3. eleven
4. have
5. It's
6. fifteen
7. What's
8. time
9. thirty
10. Can
11. tell
12. forty-five
13. do
14. have
15. o'clock

Page 109
WHAT'S THE TIME?
1. 2:00
2. 3:30
3. 4:45
4. 5:15
5. 6:00
6. 7:45

Page 110
EDITING MIX-UP
4
2
5
7
1
6
8
3

WHOSE LINE?
1. Wife
2. Husband
3. Wife
4. Wife
5. Husband
6. Husband

Page 111
YES, NO, OR MAYBE?
1. Yes
2. No
3. Maybe
4. Yes
5. No
6. Yes
7. Yes
8. Yes
9. Yes
10. Yes
11. Yes
12. Maybe
13. No

Page 112
EDITING MIX-UP
3
6
5
1
4
2
7

Page 113
FINISH THE SONG!
1. week
2. year
3. wait
4. In
5. summer
6. fall
7. waiting
8. I'm
9. February
10. April
11. July
12. September
13. December
14. It's
15. after
16. past
17. to
18. wait
19. day
20. month
21. right
22. with
23. to be
24. you

WHAT'S NEXT?
1. March
2. July
3. evening
4. Wednesday
5. winter

Page 114
WHAT ARE THEY SAYING?
1. A. What, does
   B. leaves, half past three
2. A is it
   B. a quarter to three
3. A. What time does
   B. begins, a quarter after eight
4. A. time
   B. twelve, midnight
5. A. Do
   B. It's a quarter to eight
6. A. What time, begin
   B. It, half past two

SEGMENT 24

Page 120

EDITING MIX-UP

a. 2, 1, 3
b. 3, 2, 1
c. 4, 2, 3, 1
d. 4, 2, 1, 3
e. 2, 3, 1
f. 2, 4, 1, 3

PICTURE THIS!

1. c
2. f
3. b
4. e
5. a
6. d

Page 121

THE NEXT LINE

1. b
2. a
3. b
4. a
5. a
6. b

MATCH THE SENTENCES!

1. d
2. c
3. a
4. b

Page 122

PREVIEW

1. heachache
2. backache
3. stomachache
4. cold
5. sore throat
6. earache
7. toothache

SOUND CHECK

1. a
2. b
3. b
4. b
5. b
6. a
7. a

Page 123

SOUND CHECK

1. worked
2. cooked
3. washed
4. played
5. cleaned
6. studied
7. painted
8. planted
9. waited

Page 124

EDITING MIX-UP

2, 4, 1, 6, 7, 3, 5

Page 125

SOUND CHECK

1. stomachache
2. ate
3. headache
4. drank
5. backache
6. sat
7. throat
8. sang

THE NEXT LINE

1. a
2. b
3. a
4. a
5. b
6. a
7. a
8. a

Page 126

FINISH THE RAP!

1. washed
2. floors
3. floors
4. washed
5. floors
6. cleaned
7. house
8. house
9. cleaned
10. house
11. painted
12. porch
13. porch
14. painted
15. porch
16. fixed
17. sink
18. sink
19. fixed
20. sink
21. washed
22. cleaned
23. painted
24. fixed

Page 127

WHAT'S THE RIGHT WORD?

1. cleaned, backache
2. watches
3. ate, stomachache
4. played, played
5. waited
6. talked, listened
7. studied, studied
8. drink, headache
9. cooked, cleaned
10. exercise, exercised

WHAT DID THEY DO?

1. played
2. cleaned
3. ate
4. washed
5. cooked
6. sat
7. sang
8. drank

SEGMENT 25

Page 132

SOUND CHECK

1. Did
2. did
3. Did
4. brush
5. did
6. Did
7. sleep
8. didn't
9. Did you brush
10. didn't
11. brushed
12. your hair

Page 133

WHAT DID THEY DO?

1. worked in his garden
2. visited some old friends from college
3. talked, talked, talked
4. played baseball with her kids
5. stayed home and rested
6. prepared for Leonard's visit
7. dusted the furniture
8. vacuumed the rugs
9. cleaned all the bathrooms
10. washed the windows
11. cooked
12. baked

Page 134

YES, NO, OR MAYBE?

1. Yes
2. No
3. Maybe
4. No
5. Yes
6. Maybe
7. Maybe
8. Maybe

Page 135

WHERE DID THEY GO?

		✓
	✓	✓
✓	✓	✓

SOUND CHECK

1. Did
2. go
3. didn't
4. went
5. did
6. go
7. this
8. didn't
9. to the
10. go bowling
11. This
12. went
13. went
14. to the library
15. went
16. laundromat
17. go
18. went

Page 136

YES, NO, OR MAYBE?
1. Yes
2. No
3. No
4. Maybe
5. No
6. No
7. Maybe
8. Yes
9. Yes
10. Maybe
11. Yes
12. Yes
13. Maybe
14. Yes
15. Yes
16. Maybe
17. No
18. Yes
19. Yes
20. No
21. No
22. Yes

Page 137

SOUND CHECK
1. had
2. wrote
3. read
4. worked
5. left
6. missed
7. had
8. have
9. watched
10. rested
11. took
12. read
13. drank
14. went

WHAT'S MY LINE?
1. Did, take — took
2. Did, buy — bought
3. Did, get up — got up
4. Did, have — had
5. Did, write — wrote

Page 138

SOUND CHECK
1. bought
2. were
3. was
4. wife
5. children
6. tired
7. we're
8. bought

EDITING MIX-UP
3
5
7
8
2
9
1
10
11
4
6

Page 139

YES OR NO?
1. No
2. Yes
3. Yes
4. Yes
5. Yes
6. No
7. Yes
8. Yes
9. No
10. Yes
11. Yes
12. No
13. No
14. No
15. Yes

EDITING MIX-UP 1
2
4
5
1
3

Page 140

EDITING MIX-UP 2
3
6
4
1
7
2
5

FIX THE COMMERCIALS!
1. Pet Food
2. Toothpaste
3. Shampoo
4. Paint
5. Floor Wax
6. Sofa
7. Vitamins

Page 141

BAD WEATHER
1. He didn't go to the beach. He went to the movies.
2. She didn't work in the yard. She wrote letters.
3. She didn't wash the car. She studied English.
4. He didn't have a picnic. He ate at a restaurant.
5. She didn't walk to work. She took the bus.
6. He didn't paint the garage. He did his homework.
7. He didn't plant flowers. He read a novel.
8. She didn't play golf. She went bowling.

Page 142

TV CROSSWORD
See page 166.

SEGMENT 26

Page 148

WHERE WAS EVERYBODY?
1. was, movies
2. was, meeting
3. was, computer class
4. were, birthday party
5. were, ballgame

SOUND CHECK
1. wasn't
2. wasn't
3. weren't
4. wasn't
5. was
6. was
7. were

THE NEXT LINE
1. b
2. b
3. b
4. a
5. b
6. b

Page 149

SOUND CHECK
1. Did
2. did
3. was
4. Did
5. didn't
6. wasn't
7. Did
8. did
9. were
10. Did
11. didn't
12. weren't

Pages 150–151

ON CAMERA
1. were
2. was
3. were
4. was
5. was
6. were
7. was
8. were
9. were
10. wasn't
11. were
12. were
13. were
14. wasn't
15. was
16. were
17. was
18. weren't
19. were
20. was
21. were
22. were
23. was
24. was
25. went
26. didn't
27. went
28. didn't
29. go
30. went
31. went
32. sat

33. talked
34. didn't
35. go
36. didn't
37. sit
38. went
39. looked
40. was
41. was
42. was
43. went
44. did
45. sat
46. talked
47. ate
48. had
49. did
50. was
51. started
52. rain
53. didn't
54. rain
55. was
56. wasn't
57. wasn't
58. rained
59. didn't
60. rain
61. did
62. was
63. was
64. was

Page 152

WHAT'S THE RESPONSE?

1. j
2. f
3. h
4. g
5. d
6. c
7. i
8. b
9. a
10. e

WHAT'S THE WORD?

1. was
2. Did
3. weren't
4. were
5. was, didn't, wasn't
6. weren't
7. didn't, was
8. was, wasn't
9. didn't, was
10. were, weren't, didn't, were, Were, was

Page 153

MATCH THE LINES!

1. f
2. d
3. h
4. g
5. a
6. e
7. c
8. b

SCRAMBLED SOUND TRACK

1. Tommy's father was at a meeting.
2. Were you at your computer class?
3. It didn't rain. It was sunny./ It was sunny. It didn't rain.
4. Tommy's grandparents were at the ballgame.
5. We had a big dinner because we were hungry.
6. They went to the zoo and looked at the animals.
7. I didn't go to work yesterday because I was very sick.
8. We didn't eat ice cream. We ate popcorn./We didn't eat popcorn. We ate ice cream.
9. They sat on a bench in the park, and they talked for hours and hours.

Page 44

TV CROSSWORD

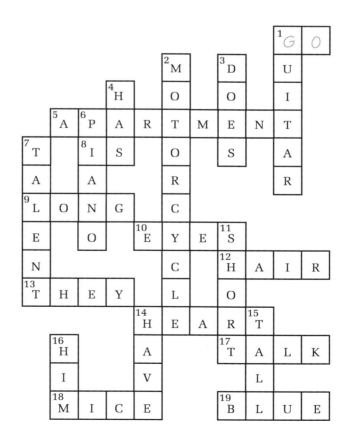

Page 142

TV CROSSWORD

Across:
1. TEETH
3. TOOK
4. VITAMINS
7. COMFORTABLE
8. TOO
9. THIS
11. SHAMPOO
13. DUSTED
15. DO
16. SAD
17. BREAKFAST

Down:
1. TO / CLEANED
2. THE
3. TOO
4. VI
5. PR
6. WAS
7. CF (FH)
8. TS (TSTE)
10. SA
14. DID
16. SO
17. BUY
18. AKATE

(Filled grid answers: TEETH, TO, THE, TOOK, TOO, VITAMINS, VI, PR, WAS, COMFORTABLE, TOO, THIS, SA, SHAMPOO, CLEANED, DO, SAD, DUSTED, DID, BREAKFAST, BUY, ATE)